Roy Ascott//Gregory Bateson//Mary Catherine Bateson//Benjamin Bogart//Pierre Bourdieu//Jack Burnham//Fritjof Capra//Paolo Cirio//Geoff Cox//Jim Crutchfield//David Dunn//Brian Eno//R. Buckminster Fuller//Frank Gillette//Michael Joaquin Grey//Boris Groys//Hans Haacke//Francis Halsall//Usman Haque// Helen Mayer Harrison//Newton Harrison//N. Katherine Hayles//Caroline A. Jones//Stephen Jones//Christian Katti//Phivos-Angelos Kollias//Bruno Latour//Mary Louise Lobsinger//Richard Paul Lohse//James Lovelock//Alessandro Ludovico//Niklas Luhmann// Humberto Maturana//Donella H. Meadows//Laurent Mignonneau//William J. Mitchell//Manfred Mohr// Nam June Paik//Gordon Pask//Philippe Pasquier// Zabet Patterson//Nick Prior//Ken Rinaldo//Claude E. Shannon//Sonia Landy Sheridan//Christa Sommerer// UBERMORGEN.COM//Francisco Varela//Steina Vasulka//Woody Vasulka//Heinz von Foerster//Warren Weaver//Michael Weinstock//Mitchell Whitelaw// Norbert Wiener//Stephen Willats//Iannis Xenakis

Systems

Whitechapel Gallery
London
The MIT Press
Cambridge, Massachusetts

Edited by Edward A. Shanken

SYSTEMS

Documents of Contemporary Art

Co-published by Whitechapel Gallery
and The MIT Press

First published 2015

Whitechapel Gallery is the imprint of Whitechapel Gallery Ventures Limited

ISBN 978-0-85488-234-2 (Whitechapel Gallery)
ISBN 978-0-262-52719-4 (The MIT Press)

A catalogue record for this book is available from the British Library

Library of Congress Cataloging-in-Publication Data

Systems / edited by Edward A. Shanken.
pages cm. – (Whitechapel: documents of contemporary art)
Includes bibliographical references and index.
ISBN 978-0-262-52719-4 (pbk. : alk. paper)
1. Arts, Modern—20th century—Philosophy.
2. Arts, Modern—21st century—Philosophy.
3. System theory.
NX456.S875 2015
700.1'1—dc23
2014039754

10 9 8 7 6 5 4 3 2 1

Series Editor: Iwona Blazwick
Commissioning Editor: Ian Farr
Project Editor: Francesca Vinter
Design by SMITH
Allon Kaye, Justine Schuster
Printed and bound in China

Cover, Tomás Saraceno, *14 Billions (working title)* (2010). Black cords, elastic rope, hooks. Dimensions/Scale: 1:17 Latrodectus mactans web. Installation view, Bonniers Konsthall, Stockholm 2010. Photograph © Studio Tomás Saraceno, 2010.

Whitechapel Gallery Ventures Limited
77–82 Whitechapel High Street
London E1 7QX
whitechapelgallery.org
To order (UK and Europe) call +44 (0)207 522 7888
or email MailOrder@whitechapelgallery.org
Distributed to the book trade (UK and Europe only)
by Central Books
www.centralbooks.com

The MIT Press
Cambridge, MA 02142
MIT Press books may be purchased at special quantity discounts for business or sales promotional use. For information, please email special_sales@
mitpress.mit.edu

Whitechapel Gallery

Documents of Contemporary Art

In recent decades artists have progressively expanded the boundaries of art as they have sought to engage with an increasingly pluralistic environment. Teaching, curating and understanding of art and visual culture are likewise no longer grounded in traditional aesthetics but centred on significant ideas, topics and themes ranging from the everyday to the uncanny, the psychoanalytical to the political.

The Documents of Contemporary Art series emerges from this context. Each volume focuses on a specific subject or body of writing that has been of key influence in contemporary art internationally. Edited and introduced by a scholar, artist, critic or curator, each of these source books provides access to a plurality of voices and perspectives defining a significant theme or tendency.

For over a century the Whitechapel Gallery has offered a public platform for art and ideas. In the same spirit, each guest editor represents a distinct yet diverse approach – rather than one institutional position or school of thought – and has conceived each volume to address not only a professional audience but all interested readers.

We are now in transition from an object-oriented to a systems-oriented culture. Here change emanates not from things but from the way things are done

Jack Burnham, 'Systems Aesthetics', 1968

Edward A. Shanken
Introduction//Systems Thinking/Systems Art

A dream of technical control and of instant information conveyed at unthought-of velocities haunted Sixties culture. The wired, electronic outlines of a cybernetic society became apparent to the visual imagination – an immediate future ... drastically modernized by the impact of computer science. It was a technologically utopian structure of feeling, positivistic and 'scientistic'.[1]

As the epigraph above suggests, systems theory and cybernetics were not limited to science and engineering but penetrated deeply into the arts and culture. The screaming electronic feedback of Jimi Hendrix's guitar at Woodstock (1966) appropriated the US National Anthem as a counter-cultural battle cry. Steina and Woody Vasulka used all manner and combination of audio and video signals to generate electronic feedback, which they conceived of as a new artistic medium: 'We look at video feedback as electronic art material ... It's the clay, it's the air, it's the energy, it's the stone ... it's the raw material that you ... build an image with ...' (*see section 2: Cybernetic Art, Architecture and Design*)[2] Les Levine described his interactive video installation *Contact: A Cybernetic Sculpture* (1969) as '... a system that synthesizes man with his technology ... the people are the software.' The 'personality' of this 'responsive mechanism', he continued, 'reflects the attitudes of the viewers ... The circuit is open.'[3] Art critic Jack Burnham described how 'we are now in transition from an *object-oriented* to a *systems-oriented culture*. Here change emanates, not from *things*, but from *the way things are done*' (*section 3: Systems Aesthetics*). Artist and theorist Roy Ascott echoed those sentiments: 'Today we are concerned less with the essence of things than with their behaviour; not what they are but what they do. This [...] vision of our time is ultimately cybernetic' (*section 2*). These experiments and attitudes represent some early foundations of the artistic explorations of systems. But the cultural significance of cybernetics is not limited to the 1960s. Although by the late 1970s it had become so ingrained and ubiquitous as to be almost invisible, it has persisted and grown for over half a century and is still going strong. Artists and scholars continue to respond with remarkable creativity and vision to emerging fields of systems-oriented science. Like their precursors, they are grappling with and deploying successive waves of technological media and corresponding social practices in ways that expand perception and cognition. In doing so they offer far-reaching insights into the systemic interrelatedness of all things. They demonstrate, moreover, the importance of integrative thinking and artistic forms of knowledge production and critique in a global economy fuelled by the algorithmic processing of big data, in which wealth is generated by Google Adsense clicks, government surveillance is reaching Orwellian proportions, and global warming is creating extreme weather conditions that threaten cities and ecosystems.

Despite the current state of affairs, the impact of systems theory and cybernetics on all aspects of human endeavour is difficult to estimate – or overestimate. And one hopes that evolved forms of systems analysis will provide potent tools that can help correct some of the global perils resulting from an overly narrow conceptual framework that cannot see the forest for the trees. Indeed, since the mid 1990s, numerous factors (including advances in computational science, networking and visualization that facilitated the simulation of complex systems) led researchers to refocus attention explicitly on systems theory and cybernetics, generating artistic and scholarly reappraisal and further elaboration. The growing number of art exhibitions and academic publications on the topic since the mid 2000s demonstrates an ongoing fascination with its aesthetic, intellectual and scientific history, as well as its contemporary significance in understanding current problems and modelling potential futures.

Before returning to the importance of systems thinking[4] for the arts, some historical background on systems theory and cybernetics will be helpful. Biologist Ludwig von Bertalanffy first proposed general systems theory in the 1930s as an approach to understanding open systems (ones that continuously interact with their environment or surroundings). Systems theory emphasizes holism over reductionism, organism over mechanism and process over product. In contrast to traditional western scientific approaches to knowledge, it shifts attention from the absolute qualities of individual parts and addresses the organization of the whole in more relativistic terms, as a dynamic process of interaction among constituent elements. The broadly interdisciplinary field of cybernetics offered a rigorous technical foundation for systems theory, and became synonymous with it.

The first wave of cybernetics focused on how systems could maintain a steady state (homeostasis) through feedback loops, which enabled self-regulation. For example, self-regulating control mechanisms consisting of feedback loops maintain human body temperature at 98.6 degrees Fahrenheit (37 degrees Celsius). Similarly a home thermostat measures and responds to fluctuations in temperature, triggering the climate control system to either heat or cool interior air in order to maintain the desired temperature. Drawing on engineer Warren Weaver's formulation of Information Theory, cybernetics established a science of control and communication that applied to the apparent goal-seeking behaviour of both animals and machines. In this regard, cybernetics was vital to, and enriched by, early research on artificial intelligence and robotics. It supplied a theoretical model to construct and to control mechanical systems that exhibit life-like behaviours, for example the McCulloch-Pitts' artificial neuron, Grey Walter's autonomous robotic 'tortoises', and Ross Ashby's Homeostat, all of which were presented at the field's fertile breeding ground, the annual Macy Conferences held in New York between 1946 and 1953. Participants represented in this volume include Weaver and his colleague Claude Shannon, anthropologist Gregory Bateson, physicist Heinz von Foerster and mathematician Norbert Wiener. Wiener coined the term 'cybernetics' from the Greek root for 'governor' and played an important role in defining the field through his influential scientific and popular publications.

The second wave of cybernetics, championed by von Foerster, insisted on including the observer as an inextricable part of the system – a second order phenomenon – thus introducing a radical sense of reflexivity. Scientists were recognized as active participants in their own scientific experiments and inextricable from them. This observation (of the recursive nature of observation) led to the constructivist position that 'the world as we know it is our invention'. Cybernetics and systems theory fundamentally challenged conventional approaches to the production of knowledge, provoking a paradigm shift that rippled throughout all academic disciplines. These shifts in mindset seeped into popular culture through a broad range of artistic practices and exhibitions and via publications such as *Radical Software* and the *Whole Earth Catalog*, the first issue of which (1968) included a section devoted to 'Understanding Whole Systems'.

Literary critic N. Katherine Hayles identifies a third wave of cybernetics, associated with the emergent behaviour of complex systems, which focused on 'getting the system to evolve in new directions' (*section 1: Foundations*). Cellular automata theory, first proposed by John von Neumann in the 1940s, established the foundations for self-replicating, dynamical systems. Mathematician John Conway's 'Game of Life' (1970), demonstrated the potential of simple cellular automata to generate unexpectedly complex behaviour, providing a mathematical system (a universal Turing machine) capable of simulating complex systems. In 1968, artist Norman White independently demonstrated a physical computing model of cellular automata. In *First Tighten Up the Drums*, digital circuits generated complex behaviours similar to non-linear dynamic systems: a grid of lights illuminated in unpredictable patterns resembling rain dripping down a windowpane. In the 1980s, Christopher Langton used cellular automata to simulate living systems, a field of research known as artificial life (A-Life), which he hoped would enable scientists to 'locate life-as-we-know-it within the larger picture of life-as-it-could-be'.[5] Also in the 1980s, biologists Humberto Maturana and Francisco Varela extended their influential theories of autopoiesis, structural coupling and embodied cognition to draw third-order phenomena of language and society into this reflexive fold (*section 1*). As a result, the scientist – and science itself – became inextricable from the complex linguistic and cultural systems that mediate the production of knowledge and the attribution of meaning and value. As discussed below, autopoiesis, A-Life and related concepts and techniques have pervaded art practice and theory.

In 1956, artist Nicolas Schöffer explicitly introduced cybernetic concepts into his responsive sculpture *CYSP I*, the title of which is an acronym joining the first two letters of the word 'cybernetics' and 'spatiodynamic'. By 1960, Roy Ascott had begun to incorporate the ideas of cybernetics into his artistic practice, later expanding these ideas into his teaching and theoretical writing, influencing generations of artists, including students Stephen Willats (*section 2*), Brian Eno (*section 4: Generative Systems*) and Christa Sommerer and Laurent Mignonneau (*section 4*). As described by Usman Haque (*section 2*), the landmark exhibition 'Cybernetic Serendipity' at the Institute of Contemporary Arts in London (1968) included psychologist and cybernetician Gordon Pask's sculpture *Colloquy of Mobiles*, which generated complex behaviours as its elements interacted with each other and with the audience. Pask was an early innovator of human-machine interface design and conversation theory, which he integrated into educational technology, fields that remain vital research areas for artists and scientists.

In 1966 Lawrence Alloway curated 'Systemic Painting' at the Guggenheim Museum, using the term to signify a cool, non-expressionistic approach to exploring the formal possibilities of an image through repetition, as in the work of Jo Baer, Agnes Martin, Robert Ryman, and others later associated with Minimalism. A related approach characterized the 1970s Systems Group in Britain, whose modular, constructivist-inspired work was re-examined at Southampton City Art Gallery in 2008. Members included Jeffrey Steele and Malcolm Hughes, co-founders of the Experimental Program which introduced computers into the studio at London's Slade School of Fine Art in order to extend 'the scope of an idea within the terms of its original proposition'.[6] The Systems Group influenced subsequent generations of artists working with diverse techniques, including Paul Brown and Ernest Edmonds (Experimental Program alumni who use computers) and Susan Tebby (who studied with Hughes and Edmonds and uses conventional media). Similarly, James Whitney's handmade film *Yantra* (1957) looks computer generated, as do his brother John Whitney's subsequent analogue films made with a military surplus gun-controller, suggesting that digital aesthetics preceded digital art (*see Zabet Patterson in section 2*).

Burnham's *Artforum* essays 'Systems Aesthetics' (1968) and 'Real Time Systems' (1969) (*section 3*) provided a critical framework for understanding the diverse range of emerging, systems-oriented art practices, while laying a theoretical foundation that continues to inspire artists. 'Systems Aesthetics' acknowledged the historic roots of 1960s systems art in the modular work of Bauhaus master Moholy-Nagy (e.g. the 'telephone paintings' of the early 1920s) and in constructivist Victor Vasarely's 1950s proposals for mass-produced art. Equally, Burnham saw it percolating in the growing disavowal by artists Ad Reinhardt, Donald Judd, Robert Morris and Les Levine of the formalist determinism of concrete objects, which was being subsumed by a growing emphasis on 'list structures' in dynamic, open-ended and responsive artworks. As he wrote, 'information, in whatever form conveyed, becomes a viable aesthetic consideration' ('Systems Aesthetics', 34). Pointing to Hans Haacke (*section 3*) as an exemplar, Burnham explained that 'real-time systems gather and process data from environments, in time to effect future events within those environments' ('Real Time Systems', 29). Haacke's real-time systems included a variety of 'plugged' and 'unplugged' media, addressing a range of natural and social systems, as did Stephen Willats' application of systems thinking to actual social intervention (*section 2*). This lineage has continued in the institutional and sociological critiques levelled by Mark Lombardi's diagrammatic 'Narrative Structures' drawings of the 1990s, Josh On's website *They Rule* (2004), and UBERMORGEN.COM et al's 'Google Will Eat Itself' (*section 4*). Similarly, it characterizes the ecological projects by artists such as Agnes Denes, David Dunn, and Newton and Helen Mayer Harrison since the 1970s, and by Beatrice da Costa, Yolande Harris, Natalie Jeremijenko, Michael Mandiberg, Andrea Polli and Aviva Rahmani since the 1990s. Indeed, from the mid 1990s to the late 2000s, Burnham's systems aesthetics had been revisited and reinterpreted to such an extent that the phenomenon of its recuperation became the subject of historiographic research in its own right (*see my text in section 3*). Following exhibitions, symposia and publications at Tate Modern in 2005 and Whitechapel Gallery in 2007, systems aesthetics reappeared (again) in US mainstream contemporary art contexts in the fiftieth anniversary issue of *Artforum* (September 2012), in which Burnham's 1968 and 1969 essays and the 'Software' exhibition he curated (1970) were subject to another round of enthusiastic reappraisals (*see Caroline Jones in section 3*).

In the early twentieth century, precursors to systems thinking can be seen percolating in the twelve-tone technique of Arnold Schoenberg, whose *Theory of Harmony* (1922) eschews traditional aesthetic conceptions of beauty and proposes instead an elaborate system of presentation, setting the stage for serial music and other parametric and generative systems of art production, sonic and visual. Its influence can be seen in the work of John Cage, Alvin Lucier, Iannis Xenakis (*section 4*) and Brian Eno (*section 4*), all of whom were well-versed in cybernetics, and who, in turn, have influenced subsequent generations of composers, artists and architects. Lucier's 1965 'Music for Solo Performer' incorporated electro-encephalography (EEG) to create a systemic bio-feedback loop between the performer's state of mind and the sound produced. With the advent of consumer EEG headsets, one can expect emerging artists to build on the pioneering work of Lucier, David Rosenboom, Richard Teitelbaum and Nina Sobell. In addition, systems thinking may have contributed to expanding the frame of art to include the total environment, as in Cage's silent composition *4'33"* and happening *Theatre Piece N.1* (both 1952), Xenakis' design for the Philips Pavilion (1958) and his *Polytopes* (1967–73) and *Diatope* (1978), Bernhard Leitner's sound environments, and Eno's multimedia software program and installation *77 Million Paintings* (2006), presented as a large-scale projection on the Sydney Opera House (1958–73), Jørn Utzon's architectural landmark inspired by natural systems.

Systems theory was an important influence on visionary architects including Buckminster Fuller (*section 5: Environmental and Social Systems*), Cedric Price (*section 2*), Paolo Soleri and the Archigram collective. Fuller's concern with sustainable urban metastructures was an important influence on Soleri and Archigram, which shared with Price a concern with creating flexible or, to use Pask's term, 'underspecified' environments that used emerging technological media to respond cybernetically to their inhabitants. As Haque, William J. Mitchell and Michael Weinstock (*section 2*) suggest, systems theory is not an outmoded way of thinking but continues to offer architecture and design new possibilities for functional and formal invention.

This generative approach heralded by Schoenberg finds visual parallels in Richard Paul Lohse's 1944 *Continually interpenetrating range of colours based on a serial system from 1–12*. Lohse, in turn, anticipates the conceptual, systemic and generative work of artists ranging from Sol LeWitt to Sonia Landy Sheridan in the 1960s and 1970s, to more contemporary works of digital art (*see section 4*). In the lineage of White and Pask but informed by the work of Maturana and Varela, in Ken Rinaldo's *Autopoiesis*, complex systemic behaviour emerges through interactions between members of a group of robotic sculptural elements and the audience. In 1990 artist Michael Joaquin Grey in collaboration with Randy Huff programmed genetic algorithms into a supercomputer to generate life-like forms analogous to actual species. In the early 1990s, ecologist Tom Ray's Tierra project used A-Life to simulate evolution, a technique used by artists including Karl Sims, Jane Prophet, and Christa Sommerer and Laurent Mignonneau. Mitchell Whitelaw and Geoff Cox note a divide between generative art which focuses on the emergent properties and potentials of formal systems and that which focuses on the critical cultural implications of the institutionalization of software. The field of bio-art, as exemplified by the Tissue Culture and Art Project, also draws on the generative aesthetic heritage, often using biological material and laboratory techniques to create awareness of and instigate critical discourse about the implications of emerging biotech practices.

Over and above Jack Burnham's theories, systems thinking has had an impact on art criticism and theory in myriad ways. Niklas Luhmann is widely recognized as the most prominent voice in establishing an explicitly systems theoretical philosophy, including insightful studies of art as a social system. But the concept of art as a social system was already present in philosopher Arthur Danto's influential 1964 essay 'The Artworld'. Danto opened the floodgates for institutional analyses of art that focus less on the objects themselves than on the larger communities or systems of discourse in which they circulate and gain meaning and value. Pierre Bourdieu's field theory shares this general approach, offering incisive commentary on the systemic relationship between art, artists, art critics, power and capital. Nick Prior applies Bourdieu's schema to the counter-cultural phenomenon of glitch music and proposes the addition of actor network theory (ANT) in order to account for the important role of technology, absent in Bourdieu's framework but vital for the analysis of glitch. ANT is itself a highly systemic form of cultural analysis. Like cybernetics, it draws parallels between human and non-human actors, both of which can exercise agency that affects the behaviour of a social system. In his analysis of art and the relationship between scientific knowledge and humanistic knowledge, Bruno Latour, a primary theorist of ANT, provides a systemic reading of artworks and systems of interpretation. If, while reading Bourdieu, Prior and Latour (*in section 5*), one substitutes the word 'system' for 'field' or 'network', the relationship of their work to systems theory becomes clear.

Strains of systems thinking can be identified throughout history and across cultures, from the I Ching to the Mayan calendar and from Buddhism to the Kabbalah. The broad appeal of systems thinking in the 1960s dovetails with the growing popularity of eastern philosophy at that time. It is not surprising, therefore, that some leading proponents of systems thinking – including, in this volume, Ascott, Burnham and Nam June Paik in the arts and Fritjof Capra, Donella Meadows and Varela in the sciences – became deeply engaged with non-western systems of thought. As such, the scientific aspects of systems theory and their impact on and implications for art demand a broader conception of systems thinking as a cultural phenomenon. For example, Charlie Gere and Fred Turner have explored the relationship of cybernetics to 1960s counter-cultural movements and the foundations of personal computing. Gere notes that in a 1972 *Rolling Stone* article, systems thinker Stewart Brand, the founding editor of *Whole Earth Catalog*, proclaimed that 'computers were coming to the people', which he thought was 'good news, maybe the best since psychedelics'.[7]

In the aftermath of World War II, systems theory provided an alternative philosophical perspective that many thinkers hoped could avoid the disastrous effects of modern technology, emblematized by the nuclear annihilation of Hiroshima and Nagasaki. At the same time, cybernetics was applied to massive air defence systems such as SAGE (Semi-Automatic Ground Environment), the iconic cybernetic Cold War computer system represented in dystopian cinema treatments such as *Dr Strangelove* and *Fail Safe* (both 1964). Perhaps its potential for peace or war, for humanitarian good or industrial excess – and the tensions generated by those oppositions – has motivated scientists and artists to emphasize the validity of systems theory for cultural, social, ethical and ecological considerations. This response is clear in many selections in this volume. Capra espouses the inevitability of an ethical sensibility resulting from systems thinking. Meadows offers a remarkable set of systems-thinking maxims to live by, culled through experience. Similarly, Bateson had recourse to the kind of 'wisdom' he associated with art and claimed that in 'a world of circuit structures' art can correct 'a too purposive view of life and mak[e] the view more systemic'. The epistemic challenges posed by second-order cybernetics pushed von Foerster to become as much a philosopher as a scientist, corroborating Latour's claim that 'no discipline is the final arbiter of any other'. Indeed, the fundamentally interdisciplinary nature of cybernetics from its inception, and the ongoing insistence of many of its practitioners on the vital importance of working across fields, suggests that the silo mentality of individual disciplines itself is a major hindrance and that the complex problems of our time can only be solved through transdisciplinary research, such as that of sound artist and composer David Dunn and the mathematician and physicist Jim Crutchfield (*section 5*). As Werner Heisenberg observed, 'the most fruitful developments frequently take place at those points where two different lines of thought meet.' Perhaps contemporary forms of systems thinking can serve as the common ground that enables committed thinkers and doers from diverse backgrounds and perspectives to integrate their ideas and methods in synergetic forms of cultural practice that spark new forms of creativity and innovation – innovation not just as the 'next big thing' in Silicon Valley but as constituting more subtle and perhaps more insidious and profound shifts in the conception and construction of knowledge and society.

1 David Mellor, *The Sixties Art Scene in London* (London and New York: Phaidon Press, 1993) 107.

2 Steina and Woody Vasulka, in Jud Yalkut, *Electronic Zen* (1973), unpublished manuscript, 28–30.

3 Gene Youngblood, *Expanded Cinema* (New York: Dutton, 1970) 340.

4 In contrast with the scientific terms 'systems theory' and 'cybernetics' I use the term 'systems thinking' to refer more generally to the related epistemological frame or mindset.

5 Christopher Langton, 'Artificial Life' (1989), in Langton, ed., *Artificial Life* (Redwood City, California: Addison-Wesley, 1992) 1.

6 Ernest Edmonds, quoting Jeffrey Steele (1967), in introduction, *Automatic Art: Human and Machine Processes that Make Art* (London: GV Art, 2014) 4.

7 Charlie Gere, *Digital Culture* (London: Reaktion, 2002; 2nd edition, 2009) 129.

CONSTRUCTIVE, LOGICAL, SYSTEMATIC OR SERIAL ART IS A SUBLIMATED AND CRITICAL ECHO OF THE STRUCTURES OF CIVILIZATION

Richard Paul Lohse, 'Lines of Development', 1943–84

FOUNDATIONS

Fritjof Capra
Systems Theory and the New Paradigm//1988

The dramatic change in concepts and ideas that happened in physics during the first three decades of [the twentieth] century [...] led Thomas Kuhn to the notion of a scientific paradigm, a constellation of [...] concepts, values, techniques, and so on, shared by a scientific community and used by that community to define legitimate problems and solutions. Changes of paradigms, according to Kuhn, occur in discontinuous, revolutionary breaks called paradigm shifts.

Today [...] we recognize paradigm shifts in physics as an integral part of a much larger cultural transformation. The intellectual crisis of quantum physicists in the 1920s is mirrored today by a similar but much broader cultural crisis. The major problems of our time – the growing threat of nuclear war, terrorism, the devastation of our natural environment, our inability to deal with poverty and starvation around the world, to name just the most urgent ones – are all different facets of one single crisis, which is essentially a crisis of perception. Like the crisis in quantum physics, it derives from the fact that most of us, and especially our large social institutions, subscribe to the concepts of an outdated worldview, inadequate for dealing with the problems of overpopulated, globally interconnected world. At the same time, researchers in several scientific disciplines, various social movements and numerous alternative organizations and networks are developing a new vision of reality that will form the basis of our future technologies, economic systems and social institutions.

What we are seeing today is a shift of paradigms not only within science but also in the larger social arena. To analyse that cultural transformation, I have generalized Kuhn's account of a scientific paradigm to that of a *social paradigm*, which I define as 'a constellation of concepts, values, perceptions and practices shared by a community, which form a particular vision of reality that is the basis of the way the community organizes itself'.[1]

The social paradigm [...] consists of a number of ideas and values, among them the view of the universe as a mechanical system composed of elementary building blocks, the view of the human body as a machine, the view of life in a society as a competitive struggle for existence, the belief in unlimited material progress to be achieved through economic and technological growth and, last but not least, the belief that a society, in which the female is everywhere subsumed under the male, is one that follows from some basic law of nature. During recent decades, all of these assumptions have been found severely limited and in need of radical revision. [...]

The emerging new paradigm may be called a holistic, or an *ecological* worldview, using the term ecological here in a much broader and deeper sense. [...] Ecological awareness, in that deep sense, recognizes the fundamental interdependence of all phenomena and the embeddedness of individuals and societies in the cyclical processes of nature.

Ultimately, deep ecological awareness is spiritual or religious awareness. When the concept of the human spirit is understood as the mode of consciousness in which the individual feels connected to the cosmos as a whole, which is the root meaning of the word *religion* (from the Latin *religare*, meaning 'to bind strongly'), it becomes clear that ecological awareness is spiritual in its deepest essence. It is, therefore, not surprising that the emerging new vision of reality, based on deep ecological awareness, is consistent with the 'perennial philosophy' of spiritual traditions, for example, that of Eastern spiritual traditions, the spirituality of Christian mystics, or with the philosophy and cosmology underlying the Native American traditions.[2]

The Systems Approach

In science, the language of systems theory, and especially the theory of living systems, seems to provide the most appropriate formulation of the new ecological paradigm.[3] Since living systems cover such a wide range of phenomena – individual organisms, social systems and ecosystems – the theory provides a common framework and language for biology, psychology, medicine, economics, ecology and many other sciences, a framework in which the so urgently needed ecological perspective is explicitly manifest. [...]

1. *Shift from the part to the whole*. In the old paradigm, it is believed that in any complex system the dynamics of the whole can be understood from the properties of the parts. The parts themselves cannot be analysed any further, except by reducing them to still smaller parts. Indeed, physics has been progressing in that way, and at each step there has been a level of fundamental constituents that could not be analysed any further.

In the new paradigm, the relationship between the parts and the whole is reversed. The properties of the parts can be understood only from the dynamics of the whole. In fact, ultimately there are no parts at all. What we call a part is merely a pattern in an inseparable web of relationships. [...]

2. *Shift from structure to process*. In the old paradigm, there are fundamental structures, and then there are forces and mechanisms through which these interact, thus giving rise to processes. In the new paradigm, every structure is seen as the manifestation of an underlying process. The entire web of relations is intrinsically dynamic. The shift from structure to process is evident, for example, when we remember that mass in contemporary physics is no longer seen as

measuring a fundamental substance but rather as a form of energy, that is, as measuring activity [...]

3. *Shift from objective to 'epistemic' science.* In the old paradigm, scientific descriptions are believed to be objective, that is, independent of the human observer and the process of knowing. In the new paradigm, it is believed that epistemology – the understanding of the process of knowledge – has to be included explicitly in the description of natural phenomena. This recognition entered into physics with Heisenberg and is closely related to the view of physical reality as a web of relationships. Whenever we isolate a pattern in this network and define it as a part, or an object, we do so by cutting through some of its connections to the rest of the network, and this may be done in different ways. As Heisenberg put it, 'What we observe is not nature itself, but nature exposed to our method of questioning.'[4]

This method of questioning, in other words epistemology, inevitably becomes part of the theory. [...]

4. *Shift from 'building' to 'network' as metaphor of knowledge.* The metaphor of knowledge as a building has been used in Western science and philosophy for thousands of years. There are fundamental laws, fundamental principles, basic building blocks, and so on. The edifice of science must be built on firm foundations. During periods of paradigm shift, it was always felt that the foundations of knowledge were shifting, or even crumbling, and that feeling induced great anxiety. [...]

In the new paradigm, the metaphor of knowledge as a building is being replaced by that of the network. Since we perceive reality as a network of relationships, our descriptions, too, form an interconnected network of concepts and models in which there are no foundations. For most scientists this metaphor of knowledge as a network with no firm foundations is extremely uncomfortable. It is explicitly expressed in physics in Geoffrey Chew's bootstrap theory of particles.[5] According to Chew, nature cannot be reduced to any fundamental entities but has to be understood entirely through self-consistency. There are no fundamental equations or fundamental symmetries in the bootstrap theory. Physical reality is seen as a dynamic web of interrelated events. Things exist by virtue of their mutually consistent relationships, and all of physics has to follow uniquely from the requirement that its components be consistent with one another and themselves. This approach is so foreign to our traditional scientific ways of thinking that it is pursued today only by a small minority of physicists. [...]

5. *Shift from truth to approximate descriptions.* The four criteria of systems thinking presented so far are all interdependent. Nature is seen as an interconnected, dynamic web of relationships, in which the identification of specific patterns as 'objects' depends on the human observer and the process of knowledge. This web

of relationships is described in terms of a corresponding network of concepts and models, none of which is any more fundamental than the others.

This new approach immediately raises an important question: If everything is connected to everything else, how can you ever hope to understand anything? Since all natural phenomena are ultimately interconnected, in order to explain any one of them we need to understand all the others, which is obviously impossible.

What makes it possible to turn the systems approach into a scientific theory is the fact that there is such a thing as approximate knowledge. [...] [I]t is recognized that all scientific concepts and theories are limited and approximate. Science can never provide any complete and definitive understanding. Scientists do not deal with truth in the sense of a precise correspondence between the description and the described phenomena. They deal with limited and approximate descriptions of reality. Heisenberg often pointed out that important fact. For example, he wrote in *Physics and Philosophy*, 'The often discussed lesson that has been learned from modern physics [is] that every word or concept, clear as it may seem to be, has only a limited range of applicability.'[6]

Self-Organizing Systems

The broadest implications of the systems approach are found today in a new theory of living systems, which originated in cybernetics in the 1940s and emerged in its main outlines over the last [forty] years.[7] [...]

The central concept of the new theory is that of self-organization. A living system is defined as a self-organizing system, which means that its order is not imposed by the environment but is established by the system itself. In other words, self-organizing systems exhibit a certain degree of autonomy. This does not mean that living systems are isolated from their environment; on the contrary, they interact with it continually, but this interaction does not determine their organization.

In this essay, I can give only a brief sketch of the theory of self-organizing systems. To do so, let me distinguish three aspects of self-organization:

1. *Pattern of organization*: the totality of relationships that define the system as an integrated whole

2. *Structure*: the physical realization of the pattern of organization in space and time

3. *Organizing activity*: the activity involved in realizing the pattern of organization

For self-organizing systems, the pattern of organization is characterized by a mutual dependency of the system's parts, which is necessary and sufficient to understand the parts. [...]

The pattern of self-organization has been studied extensively and described precisely by Humberto Maturana and Francisco Varela, who have called it *autopoiesis*, which means literally self-production. [...]

An important aspect of the theory is the fact that the description of the pattern of self-organization does not use any physical parameters, such as energy or entropy, nor does it use the concepts of space and time. [...] This pattern can be realized in space and time in different physical structures, which are then described in terms of the concepts of physics and chemistry. But such a description alone will fail to capture the biological phenomenon of self-organization. [...]

The structure of self-organizing systems has been studied extensively by Ilya Prigogine, who has called it a dissipative structure.[8] The two main characteristics of a dissipative structure are (1) that it is an open system, maintaining its pattern of organization through continuous exchange of energy and matter with its environment; and (2) that it operates far from thermodynamic equilibrium and thus cannot be described in terms of classical thermodynamics. [...]

The organizing activity of living, self-organizing systems, finally, is cognition, or mental activity. This implies a radically new concept of mind, which was first proposed by Gregory Bateson.[9] Mental process is defined as the organizing activity of life. This means that all interactions of a living system with its environment are cognitive, or mental interactions. With this new concept of mind, life and cognition become inseparably connected. Mind, or more accurately, mental process is seen as being immanent in matter at all levels of life. [...]

Science and Ethics

A further reason why I find the theory of self-organizing systems so important is that it seems to provide the ideal scientific framework for an ecologically oriented ethics.[10] Such a system of ethics is urgently needed, since most of what scientists are doing today is not life-furthering and life-preserving but life-destroying. [...]

It is generally not recognized in our culture that values are not peripheral to science and technology but constitute their very basis and driving force. During the scientific revolution in the seventeenth century, values were separated from facts, and since that time we have tended to believe that scientific facts are independent of what we do and, therefore, independent of our values. In reality, scientific facts emerge out of an entire constellation of human perceptions, values and actions – in a word, out of a paradigm – from which they cannot be separated. [...] Scientists, therefore, are responsible for their research not only intellectually but also morally.

One of the most important insights of the new systems theory of life is that life and cognition are inseparable. The process of knowledge is also the process of self-organization, that is, the process of life. Our conventional model of knowledge is one of a representation or an image of independently existing facts, which is the model derived from classical physics. From the new systems point of view, knowledge is part of the process of life, of a dialogue between object and subject.[...]

1 [footnote 3 in source] Fritjof Capra, 'The Concept of Paradigm and Paradigm Shift' and 'New Paradigm Thinking in Science', *ReVision*, vol. 9, no. 1 (Summer/Fall 1986) 11.
2 [4] Fritjof Capra, *The Tao of Physics*, 2nd edition (New York: Bantam Books, 1984).
3 [5] Fritjof Capra, *The Turning Point* (New York: Bantam Books, 1983).
4 [9] Werner Heisenberg, *Physics and Philosophy* (New York: Harper and Row, 1971) 58.
5 [11] See Fritjof Capra, 'Bootstrap Physics: A Conversation with Geoffrey Chew', in *A Passion for Physics: Essays in Honour of Geoffrey Chew*, including an interview with Chew, ed. C. De Tar, J. Finkelstein and Chung-I Tung (Philadelphia: World Scientific, 1985) 247–86.
6 [12] Werner Heisenberg, *Physics and Philosophy*, op. cit., 125.
7 [13] Fritjof Capra, *The Turning Point*, op. cit., chapter 9.
8 [15] Ilya Prigogine, *From Being to Becoming* (San Francisco: W.H. Freeman, 1980).
9 [16] Gregory Bateson, *Mind and Nature* (New York: Dutton, 1979).
10 [17] Fritjof Capra, ed., 'Science and Ethics', Elmwood Discussion transcript no. l, Elmwood Institute, Berkeley, California.

Fritjof Capra, extracts from 'Physics and the Current Change of Paradigms', edited paper from September 1986 conference at Colorado State University, in *The World View of Contemporary Physics: Does it Need a New Metaphysics?*, ed. Richard F. Kitchener (Albany: State University of New York Press, 1988) 144–52.

Heinz von Foerster
To Know and to Let Know: An Applied Theory of Knowledge//1979

[...] How does one recognize a constructivist? Very easily. If you were to ask one whether something, say, a formula, a notion, an object, order, symmetry, a taxonomy, laws of nature, etc., etc., is discovered or invented, a constructivist would tend to say invented. Moreover, if hard pressed, a constructivist would even say that the world as we know it is our invention. Since whatever we invent is our responsibility, the constructivist position contains the seed for an ethic.

I realize that I might not easily get away with such far out propositions. I will, therefore, muster whatever help I can get. [...] Let me read a charming vignette written by Gregory Bateson. He packed a lot of epistemology into a minimal space by using the literary device of a dialogue between a precocious daughter and her father. He called them 'Metalogues'. I shall give you, along with some of my comments, the one entitled 'Metalogue: what is an instinct?'

Daughter: Daddy, what is an instinct?

Let me interrupt by asking you to stop and think how you would have answered your daughter's (or son's) question. I would have proudly come up with a lexical definition: 'An instinct, my dear, is the innate aspect of behaviour that is unlearned, complex, etc., etc. …' Since the daughter could have found this kind of answer in any dictionary, her father reframes the context of the question by ignoring the semantic significance of the word 'instinct' and shifts to its functional (even political!) significance when used by one partner in a dialogue:

Father: An instinct, my dear, is an explanatory principle.

Let me pause again and invite you to reflect on the question of whether a library could accommodate the contextual switch demonstrated by the father. I consider this transition from a monological to a dialogical situation of the greatest importance, and I shall return to this later. Now let us hear what the daughter has to say to this answer.

D: But what does it explain?

F: Anything, almost anything at all. Anything you want it to explain.

Please note that something that explains almost anything at all, most likely explains nothing at all. The daughter senses this:

D: Don't be silly. It doesn't explain gravity.

F: No, but that is because nobody wants instinct to explain gravity. If they did it would explain it. We would simply say that the moon has an instinct whose strength varies inversely as the square of the distance, and so on and so on.

D: But that's nonsense, Daddy.

F: Yes, surely, but it was you who mentioned instinct, not I.

I shall not interrupt the dynamics of this dialogue any more but I ask you to pay attention to father's consistent reference to descriptions of observations and not to the observations per se (e.g. '… if you say … there was a full moon …' and not '… if there was a full moon…', etc.). Most likely you [the audience], as librarians, would have caught this anyway. Well, here we go.

D: But what does explain gravity?

F: Nothing, my dear, because gravity is an explanatory principle.

D: Oh, do you mean that you cannot use one explanatory principle to explain another – never?

F: Hum, haw, hardly ever. That is what Newton meant when he said *hypothesis non fingo*.

D: And what does that mean please?

F: Well, you know what hypotheses are. Any statement linking together two descriptive statements is an hypothesis. If you say there was a full moon on 1 February and another on 1 March, and then you link these two observations together in any way, the statement which links them is an hypothesis.

D: Yes, and I know what *non* means, but what is *fingo*?

F: Well, *fingo* is a Latin word for 'to make'. It forms a verbal noun, *fictio*, from which we get the word fiction.

D: Daddy, do you mean that Sir Isaac Newton thought that all hypotheses are just made up like stories?

F: Yes, precisely that.

D: But didn't he discover gravity? With the apple?

F: No, my dear, he invented it.

The dialogue continues, but I shall stop here because I just wanted you to hear this punchline.

Constructivists would insist that not only do we invent the laws of nature, we construct our own realities. [...]

Heinz von Foerster, extract from 'To Know and to Let Know: An Applied Theory of Knowledge', in *Communication and Control*, ed. Klaus Krippendorff (New York: Gordon and Breach, 1979); reprinted in *Canadian Library Journal*, vol. 39, no. 5 (October 1982) 283–6.

Claude E. Shannon
A Mathematical Theory of Communication//1948

[...] The fundamental problem of communication is that of reproducing at one point, either exactly or approximately, a message selected at another point. Frequently the messages have *meaning*; that is they refer to, or are correlated according to, some system with certain physical or conceptual entities. These semantic aspects of communication are irrelevant to the engineering problem. The significant aspect is that the actual message is one *selected from a set* of possible messages. The system must be designed to operate for each possible selection, not just the one which will actually be chosen, since this is unknown at the time of design.

If the number of messages in the set is finite then this number, or any monotonic function of this number, can be regarded as a measure of the information produced when one message is chosen from the set, all choices being equally likely. [...]

By a communication system we will mean a system [which] consists of essentially five parts:

1. An *information source*, which produces a message or sequence of messages to be communicated to the receiving terminal. The message may be of various types: (a) A sequence of letters, as in a telegraph or teletype system; (b) A single function of time $f(t)$, as in radio or telephony; (c) A function of time and other variables, as in black and white television [...] (d) Two or more functions of time, [as in] 'three-dimensional' sound transmission, or if the system is intended to service several individual channels in multiplex; (e) Several functions of several variables, [as in] colour television [...]; (f) Various combinations also occur, for example in television with an associated audio channel.

2. A *transmitter*, which operates on the message in some way to produce a signal suitable for transmission over the channel. In telephony this operation consists merely of changing sound pressure into a proportional electrical current. [...] Vocoder systems [analysing and synthesizing human speech], television and frequency modulation are other examples of complex operations applied to the message to obtain the signal.

3. The *channel* is merely the medium used to transmit the signal from transmitter to receiver. It may be a pair of wires, a coaxial cable, a band of radio frequencies, a beam of light, etc.

4. The *receiver* ordinarily performs the inverse operation of that done by the transmitter, reconstructing the message from the signal.

5. The *destination* is the person (or thing) for whom the message is intended.

We wish to consider certain general problems involving communication systems. To do this it is first necessary to represent the various elements involved as mathematical entities, suitably idealized from their physical counterparts. We may roughly classify communication systems into three main categories: *discrete*, *continuous* and *mixed*. By a discrete system we will mean one in which both the message and the signal are a sequence of discrete symbols. A typical case is telegraphy, where the message is a sequence of letters and the signal a sequence of dots, dashes and spaces. A continuous system is one in which the message and signal are both treated as continuous functions, e.g. radio or television. A mixed system is one in which both discrete and continuous variables appear, e.g. PCM [pulse-code modulation] transmission of speech. [...]

Claude E. Shannon, extracts from 'A Mathematical Theory of Communication' (revised version of text first published in *The Bell System Technical Journal*, July–October 1948), in Claude E. Shannon and Warren Weaver, eds, *The Mathematical Theory of Communication* (Champaign, Illinois: University of Illinois Press, 1949) 31–5.

Warren Weaver
Recent Contributions to the Mathematical Theory of Communication//1949

[...] The word *communication* will be used here in a very broad sense to include all of the procedures by which one mind may affect another. This, of course, involves not only written and oral speech, but also music, the pictorial arts, the theatre, the ballet, and in fact all human behaviour. In some connections it may be desirable to use a still broader definition of communication, namely, one which would include the procedures by means of which one mechanism (say automatic equipment to track an airplane and to compute its probable future positions) affects another mechanism (say a guided missile chasing this airplane). The language of this memorandum will often appear to refer to the special, but still very broad and important, field of the communication of speech; but practically everything said applies equally well to music of any sort, and to still or moving pictures, as in television.

Three Levels of Communications Problems

Relative to the broad subject of communication, there seem to be problems at three levels. Thus it seems reasonable to ask, serially:

Level A. How accurately can the symbols of communication be transmitted? (The technical problem.)

Level B. How precisely do the transmitted symbols convey the desired meaning? (The semantic problem.)

Level C. How effectively does the received meaning affect conduct in the desired way? (The effectiveness problem.)

The *technical problems* are concerned with the accuracy of transference from sender to receiver of sets of symbols (written speech), or of a continuously varying signal (telephonic or radio transmission of voice or music), or of a continuously varying two-dimensional pattern (television), etc. Mathematically, the first involves transmission of a finite set of discrete symbols, the second the transmission of one continuous function of time, and the third the transmission of many continuous functions of time or of one continuous function of time and of two space coordinates.

The *semantic problems* are concerned with the identity, or satisfactorily close approximation, in the interpretation of meaning by the receiver, as compared with the intended meaning of the sender. This is a very deep and involved situation, even when one deals only with the relatively simpler problems of communicating through speech.

One essential complication is illustrated by the remark that if Mr X is suspected not to understand what Mr Y says, then it is theoretically not possible, by having Mr Y do nothing but talk further with Mr X, completely to clarify this situation in any finite time. If Mr Y says 'Do you now understand me?' and Mr X says 'Certainly, I do', this is not necessarily a certification that understanding has been achieved. It may just be that Mr X did not understand the question. If this sounds silly, try it again as 'Czy pafi mnie rozumie?' with the answer 'Hai wakkate imasu.' I think that this basic difficulty is, at least in the restricted field of speech communication, reduced to a tolerable size (but never completely eliminated) by 'explanations' which (a) are presumably never more than approximations to the ideas being explained, but which (b) are understandable since they are phrased in language which has previously been made reasonably clear by operational means. For example, it does not take long to make the symbol for 'yes' in any language operationally understandable.

The semantic problem has wide ramifications if one thinks of communication in general. Consider, for example, the meaning [in 1949] to a Russian of a US newsreel picture.

The *effectiveness problems* are concerned with the success with which the meaning conveyed to the receiver leads to the desired conduct on his part. It may seem at first glance undesirably narrow to imply that the purpose of all communication is to influence the conduct of the receiver. But with any reasonably broad definition of conduct, it is clear that communication either affects conduct or is without any discernible and probable effect at all.

The problem of effectiveness involves aesthetic considerations in the case of the fine arts. In the case of speech, written or oral, it involves considerations, which range all the way from the mere mechanics of style, through all the psychological and emotional aspects of propaganda theory, to those value judgements which are necessary to give useful meaning to the words 'success' and 'desired' in the opening sentence of this section on effectiveness.

The effectiveness problem is closely interrelated with the semantic problem, and overlaps it in a rather vague way; and there is in fact overlap between all of the suggested categories of problems.

Comments

So stated, one would be inclined to think that Level A is a relatively superficial one, involving only the engineering details of good design of a communication system; while B and C seem to contain most if not all of the philosophical content of the general problem of communication.

The mathematical theory of the engineering aspects of communication, as developed chiefly by Claude Shannon at the Bell Telephone Laboratories, admittedly applies in the first instance only to problem A, namely, the technical problem of accuracy of transference of various types of signals from sender to receiver. But the theory has, I think, a deep significance [...] Part of the significance of the new theory comes from the fact that [...] any limitations discovered in the theory at Level A necessarily apply to levels B and C. But a larger part of the significance comes from the fact that the analysis at Level A discloses that this level overlaps the other levels more than one could possibly naively suspect. Thus the theory of Level A is, at least to a significant degree, also a theory of levels B and C. [...]

Warren Weaver, extract from 'Recent Contributions to the Mathematical Theory of Communication' (1949), in Claude E. Shannon and Warren Weaver, eds, *The Mathematical Theory of Communication* (Champaign, Illinois: University of Illinois Press, 1949) 3–6.

Norbert Wiener
The Human Use of Human Beings//1950

[...] Fantasy has always been at the service of philosophy, and Plato was not ashamed to clothe his epistemology in the metaphor of the cave. [Jacob] Bronowski, among others, has pointed out that mathematics, which most of us see as the most factual of all sciences, constitutes the most colossal metaphor imaginable, and must be judged, aesthetically as well as intellectually, in terms of the success of this metaphor.

The metaphor to which I devote this [text] is one in which the organism is seen as message. Organism is opposed to chaos, to disintegration, to death, as message is to noise. To describe an organism, we do not try to specify each molecule in it, and catalogue it bit by bit, but rather to answer certain questions about it which reveal its pattern: a pattern which is more significant and less probable as the organism becomes, so to speak, more fully an organism.

We have already seen that certain organisms, such as man, tend for a time to maintain and often even to increase the level of their organization, as a local enclave in the general stream of increasing entropy, of increasing chaos and de-differentiation. Life is an island here and now in a dying world. The process by which we living beings resist the general stream of corruption and decay is known as *homeostasis*.

We can continue to live in the very special environment which we carry forward with us only until we begin to decay more quickly than we can reconstitute ourselves. Then we die. If our bodily temperature rises or sinks one degree from its normal level of 98.6 degrees Fahrenheit, we begin to take notice of it, and if it rises or sinks ten degrees, we are all but sure to die. The oxygen and carbon dioxide and salt in our blood, the hormones flowing from our ductless glands, are all regulated by mechanisms which tend to resist any untoward changes in their levels. These mechanisms constitute what is known as homeostasis, and are negative feedback mechanisms of a type that we may find exemplified in mechanical automata.

It is the pattern maintained by this homeostasis, which is the touchstone of our personal identity. Our tissues change as we live: the food we eat and the air we breathe become flesh of our flesh and bone of our bone, and the momentary elements of our flesh and bone pass out of our body every day with our excreta. We are but whirlpools in a river of ever-flowing water. We are not stuff that abides, but patterns that perpetuate themselves.

A pattern is a message, and may be transmitted as a message. How else do we employ our radio than to transmit patterns of sound, and our television set than to transmit patterns of light? It is both amusing and instructive to consider what would happen if we were to transmit the whole pattern of the human body, of the brain with its memories and cross connections, so that a hypothetical receiving instrument could re-embody these messages in appropriate matter, capable of continuing the processes already in the body and the mind, and of maintaining the integrity needed for this continuation by a process of homeostasis [...]

If we consider the two types of communication: namely, material transport, and transport of information alone, it is at present possible for a person to go from one place to another only by the former, and not as a message. However, even now the transportation of messages serves to forward an extension of man's senses and his capabilities of action from one end of the world to another [...] [T]he distinction between material transportation and message transportation is not in any theoretical sense permanent and unbridgeable.

This takes us very deeply into the question of human individuality. The problem of the nature of human individuality and of the barrier which separates one personality from another is as old as history [...]

One thing at any rate is clear. The physical identity of an individual does not consist in the matter of which it is made. Modern methods of tagging the elements participating in metabolism have shown a much higher turnover than was long thought possible, not only of the body as a whole, but of each and every component part of it. The biological individuality of an organism seems to lie in a certain continuity of process, and in the memory by the organism of the effects of its past development. This appears to hold also of its mental development. In terms of the computing machine, the individuality of a mind lies in the retention of its earlier tapings and memories, and in its continued development along lines already laid out.

Under these conditions, just as a computing machine may be used as a pattern on which to tape other computing machines, and just as the future development of these two machines will continue parallel except for future changes in taping and experience, so too, there is no inconsistency in a living individual forking or divaricating into two individuals sharing the same past, but growing more and more different. This is what happens with identical twins; but there is no reason why it could not happen with what we call the mind, without a similar split of the body. To use computing-machine language again, at some stage a machine, which was previously assembled in an all-over manner, may find its connections divided into partial assemblies with a higher or lower degree of independence [...]

Moreover, it is thinkable that two large machines which had previously not been coupled may become coupled so as to work from that stage on as a single machine. Indeed this sort of thing occurs in the union of the germ cells, although perhaps not on what we would ordinarily call a purely mental level. [...]

To recapitulate: the individuality of the body is that of a flame rather than that of a stone, of a form rather of a bit of substance. This form can be transmitted or modified and duplicated, although at present we know only how to duplicate it over a short distance.

When one cell divides into two, or when one of the genes which carries our corporeal and mental birthright is split in order to make ready for a reduction division of a germ cell, we have a separation in matter which is conditioned by the power of a pattern of living tissue to duplicate itself. Since this is so, there is no absolute distinction between the types of transmission which we can use for sending a telegram from country to country and the types of transmission which at least are theoretically possible for transmitting a living organism such as a human being.

Let us then admit that the idea that one might conceivably travel by telegraph, in addition to travelling by train or airplane, is not intrinsically absurd, far as it may be from realization. The difficulties are, of course, enormous [...]

I have stated these things not because I want to write a science fiction story concerning itself with the possibility of telegraphing a man, but because it may help us understand that the fundamental idea of communication is that of the transmission of messages, and that the bodily transmission of matter and messages is only one conceivable way of attaining that end. [...]

Norbert Wiener, extracts from *The Human Use of Human Beings* (London: Eyre & Spottiswoode, 1950) 95–104.

N. Katherine Hayles

Contesting for the Body of Information: The Macy Conferences on Cybernetics (1946 and 1953)//1999

When and where did information get constructed as a disembodied medium? How were researchers convinced that humans and machines are brothers under the skin? Although the Macy Conferences on Cybernetics were not the only forum grappling with these questions, they were particularly important because they acted as a crossroads for the traffic in cybernetic models and artefacts. This [text] charts the arguments that made information seem more important than materiality within this research community. Broadly speaking, the arguments were deployed along three fronts. The first was concerned with the construction of information as a theoretical entity; the second, with the construction of (human) neural structures so that they were seen as flows of information; the third, with the construction of artefacts that translated information flows into observable operations, thereby making the flows 'real'.

Yet at each of these fronts, there was also significant resistance to the reification of information. Alternate models were proposed; important qualifications were voiced; objections were raised to the disparity between simple artefacts and the complex problems they addressed. Reification was triumphant not because it had no opposition but because scientifically and culturally situated debates made it seem a better choice than the alternatives. Recovering the complexities of these debates helps to demystify the assumption that information is more essential than matter or energy. Followed back to moments before it became a black box, this conclusion seems less like an inevitability and more like the result of negotiations specific to the circumstances of the US techno-scientific culture during and immediately following World War II.

The Macy Conferences were unusual in that participants did not present finished papers. Rather, speakers were invited to sketch out a few main ideas to initiate discussion. The discussions, rather than the presentations, were the centre of interest. Designed to be intellectual free-for-alls, the conferences were radically interdisciplinary. The transcripts show that researchers from a wide variety of fields – neurophysiology, electrical engineering, philosophy, semantics, literature and psychology, among others – struggled to understand one another and make connections between others' ideas and their own areas of expertise. In the process, a concept that may have begun as a model of a particular physical system came to have broader significance, acting simultaneously as mechanism and metaphor.

The dynamics of the conferences facilitated this mixing. Researchers might not have been able to identify in their own work the mechanism discussed by a fellow participant, but they could understand it metaphorically and then associate the metaphor with something applicable to their own field. The process appears repeatedly throughout the transcripts. When Claude Shannon used the word 'information', for example, he employed it as a technical term having to do with message probabilities. When Gregory Bateson appropriated the same word to talk about initiation rituals, he interpreted it metaphorically as a 'difference that makes a difference' and associated it with feedback loops between contesting social groups. As mechanism and metaphor were compounded, concepts that began with narrow definitions spread out into networks of broader significance. Earlier [in *How We Became Postmodern*] I called these networks 'constellations', suggesting that during the Macy period, the emphasis was on homeostasis. [Here I explore] the elements that came together to form the homeostasis constellation [and] also demonstrate the chain of associations that bound reflexivity together with subjectivity during the Macy period, which for many of the physical scientists was enough to relegate reflexivity to the category of 'non-science' rather than 'science'. [A]fter the Macy period ended [...] reflexivity was modified so that it could count as producing scientific knowledge during the second wave of cybernetics.

The Meaning(lessness) of Information

The triumph of information over materiality was a major theme at the first Macy Conference. John von Neumann and Norbert Wiener led the way by making clear that the important entity in the man-machine equation was information, not energy. [...] Central was how much information could flow through the system and how quickly it could move. Wiener, emphasizing the movement from energy to information, made the point explicitly: 'The fundamental idea is the message ... and the fundamental element of the message is the decision.' Decisions are important not because they produce material goods but because they produce information. Control information, and power follows.

But what counts as information? [...] Claude Shannon defined information as a probability function with no dimensions, no materiality, and no necessary connection with meaning. Although a full exposition of information theory is beyond the scope of this text, the following explanation, adapted from an account by Wiener, will give an idea of the underlying reasoning. Like Shannon, Wiener thought of information as representing a choice. More specifically, it represents a choice of one message from among a range of possible messages. Suppose there are thirty-two horses in a race, and we want to bet on Number 3. The bookie suspects the police have tapped his telephone, so he has arranged for his clients to use a code. He studied communication theory (perhaps he was in one of the summer-school classes on communication theory that Wiener taught at UCLA), and he knows that any message can be communicated through a binary code. When we call up, his voice program asks if the number falls in the range of 1 to 16. If it does, we punch the number '1'; if not, the number 'O'. We use this same code when the voice program asks if the number falls in the range of 1 to 8, then the range of 1 to 4, and next the range of 1 to 2. Now the program knows that the number must be either 3 or 4, so it says, 'If 3, press 1; if 4, press O', and a final tap communicates the number. Using these binary divisions, we need five responses to communicate our choice. [...]

Note that [information] theory is formulated entirely without reference to what information means. Only the probabilities of message elements enter into the equations. Why divorce information from meaning? Shannon and Wiener wanted information to have a stable value as it moved from one context to another. If it was tied to meaning, it would potentially have to change values every time it was embedded in a new context, because context affects meaning. Suppose, for example, you are in a windowless office and call to ask about the weather. 'It's raining', I say. On the other hand, if we are both standing on a street corner, being drenched by a downpour, this same response would have a very different meaning. In the first case, I am telling you something you don't know; in the second, I am being ironic (or perhaps moronic). An information concept that ties information to meaning would have to yield two different values for the two circumstances, even though the message ('It's raining') is the same.

To cut through this Gordian knot, Shannon and Wiener defined information so that it would be calculated as the same value regardless of the contexts in which it was embedded, which is to say, they divorced it from meaning. *In context*, this was an appropriate and sensible decision. *Taken out of context*, the definition allowed information to be conceptualized as if it were an entity that can flow unchanged between different material substrates, as when Hans Moravec envisions the information contained in a brain being downloaded into a computer. Ironically, this reification of information is enacted through the same kind of decontextualizing moves that the theory uses to define information as such. The theory decontextualizes information; Moravec decontextualizes the theory. Thus, a simplification necessitated by engineering considerations becomes an ideology in which a reified concept of information is treated as if it were fully commensurate with the complexities of human thought.

Shannon himself was meticulously careful about how he applied information theory, repeatedly stressing that information theory concerned only the efficient transmission of messages through communication channels, not what those messages mean. Although others were quick to impute larger linguistic and social implications to the theory, he resisted these attempts. Responding to a presentation by Alex Bavelas on group communication at the eighth Macy Conference, he cautioned that he did not see 'too close a connection between the notion of information as we use it in communication engineering and what you are doing here ... the problem here is not so much finding the best encoding of symbols ... but, rather, the determination of the semantic question of what to send and to whom to send it'. For Shannon, defining information as a probability function was a strategic choice that enabled him to bracket semantics. He did not want to get involved in having to consider the receiver's mindset as part of the communication system. He felt so strongly on this point that he suggested Bavelas distinguish between information in a channel and information in a human mind by characterizing the latter through 'subjective probabilities', although how these were to be defined and calculated was by no means clear.

Not everyone agreed that it was a good idea to decontextualize information. At the same time that Shannon and Wiener were forging what information would mean in a US context, Donald MacKay, a British researcher, was trying to formulate an information theory that would take meaning into account. At the seventh conference, he presented his ideas to the Macy group. The difference between his view and Shannon's can be seen in the way he bridled at Shannon's suggestion about 'subjective probabilities'. In the rhetoric of the Macy Conferences, 'objective' was associated with being scientific, whereas 'subjective' was a code word implying that one had fallen into a morass of unquantifiable feelings that might be magnificent but were certainly not science. MacKay's first move was to rescue information that affected the receiver's mindset from the 'subjective' label. He proposed that both Shannon and Bavelas were concerned with what he called 'selective information', that is, information calculated by considering the selection of message elements from a set. But selective information alone is not enough; also required is another kind of information that he called 'structural'. Structural information indicates how selective information is to be understood; it is a message about how to interpret a message – that is, it is a metacommunication.

To illustrate, say I launch into a joke and it falls flat. In that case, I may resort to telling my interlocutor, 'That's a joke.' The information content of this message, considered as selective information (measured in 'metrons'), is calculated with probability functions similar to those used in the Shannon Wiener theory. In addition, my metacomment also carries structural information (measured in 'logons'), for it indicates that the preceding message has one kind of structure rather than another (a joke instead of a serious statement). In another image MacKay liked to use, he envisioned selective information as choosing among folders in a file drawer, whereas structural information increased the number of drawers (jokes in one drawer, academic treatises in another).

Since structural information indicates how a message should be interpreted, semantics necessarily enters the picture. In sharp contrast to message probabilities, which have no connection with meaning, structural information was to be calculated through changes brought about in the receiver's mind. [...]

And how does one measure these changes? An observer looks at the mind of the person who received the message, which is to say that changes are made in the observer's mind, which in turn can also be observed and measured by someone else. The progression tends toward the infinite regress characteristic of reflexivity. Arguing for a strong correlation between the *nature* of a representation and its *effect*, MacKay's model recognized the mutual constitution of form and content, message and receiver. His model was fundamentally different from the Shannon-Wiener theory because it triangulated between reflexivity, information and meaning. In the context of the Macy Conferences, his conclusion qualified as radical: subjectivity, far from being a morass to be avoided, is precisely what enables information and meaning to be connected.

The problem was how to quantify the model. To achieve quantification, a mathematical model was needed for the changes that a message triggered in the receiver's mind. The staggering problems this presented no doubt explain why MacKay's version of information theory was not widely accepted among electrical engineers [...]

Nicolas S. Tzannes [...] pointed out that whereas Shannon and Wiener define information in terms of what it *is*, MacKay defines it in terms of what it *does*. The formulation emphasizes the reification that information undergoes in the Shannon-Wiener theory. Stripped of context, it becomes a mathematical quantity weightless as sunshine, moving in a rarefied realm of pure probability, not tied down to bodies or material instantiations. The price it pays for this universality is its divorce from representation. When information is made representational, as in MacKay's model, it is conceptualized as an action rather than a thing. Verb-like, it becomes a process that someone enacts, and thus it necessarily implies context and embodiment. The price it pays for embodiment is difficulty of quantification and loss of universality.

In the choice between what information is and what it does, we can see the rival constellations of homeostasis and reflexivity beginning to take shape. Making information a thing allies it with homeostasis, for so defined, it can be transported into any medium and maintain a stable quantitative value, reinforcing the stability that homeostasis implies. Making information an action links it with reflexivity, for then its effect on the receiver must be taken into account, and measuring this effect sets up the potential for a reflexive spiral through an infinite regress of observers. Homeostasis won in the first wave largely because it was more manageable quantitatively. Reflexivity lost because specifying and delimiting context quickly ballooned into an unmanageable project. At every point, these outcomes are tied to the historical contingencies of the situation – the definitions offered, the models proposed, the techniques available, the allies and resources mobilized by contending participants for their views. Conceptualizing information as a disembodied entity was not an arbitrary decision, but neither was it inevitable. […]

If humans are information processing machines, then they must have biological equipment enabling them to process binary code. The model constructing the human in these terms was the McCulloch-Pitts neuron. The McCulloch-Pitts neuron was the primary model through which cybernetics was seen as having 'a setting in the flesh', as Warren McCulloch put it. The problem was how to move from this stripped-down neural model to such complex issues as universals in thought, gestalts in perception, and representations of what a system cannot represent. Here the slippage between mechanism and model becomes important, for even among researchers dedicated to a hard-science approach, such as McCulloch, the tendency was to use the model metaphorically to forge connections between relatively simple neural circuits and the complexities of embodied experience. In the process, the disembodied logical form of the circuit was rhetorically transformed from being an *effect* of the model to a *cause* of the model's efficacy. This move […] made embodied reality into a blurred and messy instantiation of the clean abstractions of logical forms. […] The tension between logical form and embodiment […] displays how the construction of a weightless information was complicated when cybernetics moved into the intimate context of the body's own neural functioning. […]

N. Katherine Hayles, extracts from 'Contesting for the Body of Information: The Macy Conferences on Cybernetics', *How We Became Postmodern* (Chicago: University of Chicago Press, 1999) 50–57 [footnotes not included].

Gregory Bateson

Style, Grace and Information in Primitive Art//1967

[...] Aldous Huxley used to say that the central problem for humanity is the quest for *grace*. [...]

I argue that art is a part of man's quest for grace; sometimes his ecstasy in partial success, sometimes his rage and agony at failure.

I argue also that there are many species of grace within the major genus; and also that there are many kinds of failure and frustration and departure from grace. No doubt each culture has its characteristic species of grace towards which its artists strive, and its own species of failure.

Some cultures may foster a negative approach to this difficult integration, an avoidance of complexity by crass preference either for total consciousness or total unconsciousness. Their art is unlikely to be 'great'.

I shall argue that the problem of grace is fundamentally a problem of integration and that what is to be integrated is the diverse parts of the mind – especially those multiple levels of which one extreme is called 'consciousness' and the other the 'unconscious'. For the attainment of grace, the reasons of the heart must be integrated with the reasons of the reason. [...]

The central question is: In what form is information about psychic integration contained or coded in the work of art?

Style and Meaning

They say that 'every picture tells a story' [...] But I want precisely to avoid analysing the 'story'. [...]

I am concerned with what important psychic information is in the art object quite apart from what it may 'represent'. '*Le style c'est l'homme même*' ('The style is the man himself') (Buffon). What is implicit in style, materials, composition, rhythm, skill, and so on? [...]

The lions in Trafalgar Square could have been eagles or bulldogs and still have carried the same (or similar) messages about empire and about the cultural premises of nineteenth-century England. And yet, how different might their message have been had they been made of wood!

But representationalism as such is relevant. The extremely realistic horses and stags of Altamira are surely not about the same cultural premises as the highly conventionalized black outlines of a later period. The code whereby perceived objects or persons (or supernaturals) are transformed into wood or paint is a source of information about the artist and his culture.

It is the very rules of transformation that are of interest to me – not the message, but the code.

My goal is not instrumental. I do not want to use the transformation rules when discovered to undo the transformation or to 'decode' the message. To translate the art object into mythology and then examine the mythology would be only a neat way of dodging or negating the problem of 'what is art?'

I ask, then, not about the meaning of the encoded message but rather about the meaning of the code chosen. But still that most slippery word 'meaning' must be defined.

It will be convenient to define meaning in the most general possible way in the first instance.

'Meaning' may be regarded as an approximate synonym of pattern, redundancy, information and 'restraint', within a paradigm of the following sort:

Any aggregate of events or objects (e.g. a sequence of phonemes, a painting, or a frog, or a culture) shall be said to contain 'redundancy' or 'pattern' if the aggregate can be divided in any way by a 'slash mark', such that an observer perceiving only what is on one side of the slash mark can guess, with better than random success, what is on the other side of the slash mark. We may say that what is on one side of the slash contains information or has meaning about what is on the other side. Or, in engineer's language, the aggregate contains 'redundancy'. Or, again, from the point of view of a cybernetic observer, the information available on one side of the slash will restrain (i.e. reduce the probability of) wrong guessing. [...]

The essence and *raison d'être* of communication is the creation of redundancy, meaning, pattern, predictability, information, and/or the reduction of the random by 'restraint'.

It is, I believe, of prime importance to have a conceptual system which will force us to see the 'message' (e.g. the art object) as both itself internally patterned and itself a part of a larger patterned universe – the culture or some part of it.

The characteristics of objects of art are believed to be about, or to be partly derived from, or determined by, other characteristics of cultural and psychological systems. Our problem might therefore be oversimply represented by the diagram [Characteristic, of art object/Characteristics of rest of culture], where square brackets enclose the universe of relevance, and where the oblique stroke represents a slash across which some guessing is possible, in one direction or in both. The problem, then, is to spell out what sorts of relationships, correspondences, etc., cross or transcend this oblique stroke. [...]

Levels and Logical Types

[...] The word 'know' is not merely ambiguous in covering both *connaître* (to know through the senses, to recognize or perceive) and *savoir* (to know in the mind), but varies – actively shifts – in meaning for basic systemic reasons. That which we know through the senses can *become* knowledge in the mind. [...]

[T]here is a special form of 'knowing' which is usually regarded as adaptation rather than information. A shark is beautifully shaped for locomotion in water, but the genome of the shark surely does not contain direct information about hydrodynamics. Rather, the genome must be supposed to contain information or instructions which are the complement of hydrodynamics. Not hydrodynamics, but what hydrodynamics requires, has been built up in the shark's genome. Similarly, a migratory bird perhaps does not know the way to its destination in any of the senses outlined above, but the bird may contain the complementary instructions necessary to cause it to fly right.

'*Le coeur a ses raisons que la raison ne connaît point*' ('The heart has its reasons which reason does not at all perceive') [Pascal]. It is this – the complex layering of consciousness and unconsciousness – that creates difficulty when we try to discuss art or ritual or mythology. The matter of levels of the mind has been discussed from many points of view, at least four of which must be mentioned and woven into any scientific approach to art:

(1) Samuel Butler's insistence that the better an organism 'knows' something, the less conscious it becomes of its knowledge, i.e. there is a process whereby knowledge (or 'habit' – whether of action, perception or thought) sinks to deeper and deeper levels of the mind. This phenomenon, which is central to Zen discipline (see Herrigel, *Zen in the Art of Archery*), is also relevant to all art and all skill.

(2) Adalbert Ames' demonstrations that the conscious, three-dimensional visual images which we make of that which we see are made by processes involving mathematical premises of perspective, etc., of the use of which we are totally unconscious. Over these processes we have no voluntary control. A drawing of a chair with the perspective of Van Gogh affronts the conscious expectations and, dimly, reminds the consciousness of what had been (unconsciously) taken for granted.

(3) The Freudian (especially Fenichel's) theory of dreams as metaphors coded according to primary process. I shall consider style – neatness, boldness of contrast, etc. – as metaphoric and therefore as linked to those levels of the mind where primary process holds sway.

(4) The Freudian view of the unconscious as the cellar or cupboard to which fearful and painful memories are consigned by a process of repression. [...]

These considerations are especially relevant in any attempt to derive a theory of art or poetry. Poetry is not a sort of distorted and decorated prose, but rather prose is poetry which has been stripped down and pinned to a Procrustean bed of logic. The computer men who would programme the translation of languages sometimes forget this fact about the primary nature of language. To try to construct a machine to translate the art of one culture into the art of another would be equally silly. [...]

In the cliché system of Anglo-Saxons, it is commonly assumed that it would be somehow better if what is unconscious were made conscious. Freud, even, is said to have said, 'Where id was, there ego shall be', as though such an increase in conscious knowledge and control would be both possible and, of course, an improvement. This view is the product of an almost totally distorted epistemology and a totally distorted view of what sort of thing a man, or any other organism, is.

Of the four sorts of unconsciousness listed above, it is very clear that the first three are necessary. Consciousness, for obvious mechanical reasons, must always be limited to a rather small fraction of mental process. If useful at all, it must therefore be husbanded. The unconsciousness associated with habit is an economy both of thought and of consciousness; and the same is true of the inaccessibility of the processes of perception. The conscious organism does not require (for pragmatic purposes) to know how it perceives – only to know what it perceives. (To suggest that we might operate without a foundation in primary process would be to suggest that the human brain ought to be differently structured.) Of the four types, only the Freudian cupboard for skeletons is perhaps undesirable and could be obviated. But there may still be advantages in keeping the skeleton off the dining room table.

In truth, our life is such that its unconscious components are continuously present in all their multiple forms. It follows that in our relationships we continuously exchange messages about these unconscious materials, and it becomes important also to exchange meta-messages by which we tell each other what order and species of unconsciousness (or consciousness) attaches to our messages.

In a merely pragmatic way, this is important because the orders of truth are different for different sorts of messages. In so far as a message is conscious and voluntary, it could be deceitful. I can tell you that the cat is on the mat when in fact she is not there. I can tell you 'I love you' when in fact I do not. But discourse about relationship is commonly accompanied by a mass of semi-voluntary kinesic and autonomic signals which provide a more trustworthy comment on the verbal message.

Similarly with skill, the fact of skill indicates the presence of large unconscious components in the performance.

It thus becomes relevant to look at any work of art with the question: What components of this message material had what orders of unconsciousness (or consciousness) for the artist? And this question, I believe, the sensitive critic usually asks, though perhaps not consciously.

Art becomes, in this sense, an exercise in communicating about the species of unconsciousness. Or, if you prefer it, a sort of play behaviour whose function is, amongst other things, to practice and make more perfect communication of this kind. [...]

'If I could tell you what it meant, there would be no point in dancing it.' [...]

Isadora Duncan's remark [suggests that] if the message were the sort of message that could be communicated in words, there would be no point in dancing it, but it is not that sort of message. It is, in fact, precisely the sort of message that would be falsified if communicated in words, because the use of words (other than poetry) would imply that this is a fully conscious and voluntary message, and this would be simply untrue.

I believe that what Isadora Duncan or any artist is trying to communicate is more like: 'This is a particular sort of partly unconscious message. Let us engage in this particular sort of partly unconscious communication.' Or perhaps: 'This is a message about the interface between conscious and unconscious.'

The message of *skill* of any sort must always be of this kind. The sensations and qualities of skill can never be put in words, and yet the fact of skill is conscious.

The artist's dilemma is of a peculiar sort. He must practise in order to perform the craft components of his job. But to practise has always a double effect. It makes him, on the one hand, more able to do whatever it is he is attempting; and, on the other hand, by the phenomenon of habit formation, it makes him less aware of how he does it.

If his attempt is to communicate about the unconscious components of his performance, then it follows that he is on a sort of moving stairway (or escalator) about whose position he is trying to communicate but whose movement is itself a function of his efforts to communicate. Clearly, his task is impossible, but, as has been remarked, some people do it very prettily. [...]

The Corrective Nature of Art

It was noted above that consciousness is necessarily selective and partial, i.e. that the content of consciousness is, at best, a small part of truth about the self. But if this part be *selected* in any systematic manner, it is certain that the partial truths of consciousness will be, in aggregate, a distortion of the truth of some larger whole.

In the case of an iceberg, we may guess, from what is above surface, what sort of stuff is below; but we cannot make the same sort of extrapolation from the content of consciousness. It is not merely the selectivity of preference, whereby the skeletons accumulate in the Freudian unconscious, that makes such extrapolation unsound. Such a selection by preference would only promote optimism.

What is serious is the cross-cutting of the circuitry of the mind. If, as we must believe, the total mind is an integrated network (of propositions, images, processes, neural pathology, or what have you – according to what scientific language you prefer to use), and if the content of consciousness is only a sampling of different parts and localities in this network; then, inevitably, the conscious view of the network as a whole is a monstrous denial of the integration of that whole. From the cutting of consciousness, what appears above the surface is arcs of circuits instead of either the complete circuits or the larger complete circuits of circuits.

What the unaided consciousness (unaided by art, dreams, and the like) can never appreciate is the *systemic* nature of mind. [...]

Characteristically, errors occur wherever the altered causal chain is part of some large or small circuit structure of system. And the remainder of our technology [...] bids fair to disrupt the rest of our ecology. [...]

The point [...] is that mere purposive rationality unaided by such phenomena as art, religion, dream, and the like, is necessarily pathogenic and destructive of life; and that its virulence springs specifically from the circumstance that life depends upon interlocking circuits of contingency, while consciousness can see only such short arcs of such circuits as human purpose may direct.

In a word, the unaided consciousness must always involve man in the sort of stupidity of which evolution was guilty when she urged upon the dinosaurs the common-sense values of an armaments race. She inevitably realized her mistake a million years later and wiped them out. [...]

That is the sort of world we live in – a world of circuit structures – and love can survive only if wisdom (i.e. a sense or recognition of the fact of circuitry) has an effective voice. [...]

[I]f art, as suggested above, has a positive function in maintaining what I called 'wisdom', i.e. in correcting a too purposive view of life and making the view more systemic, then the question to be asked of the given work of art becomes: What sorts of correction in the direction of wisdom would be achieved by creating or viewing this work of art?

The question becomes dynamic rather than static.

Composition

[...] It is probably an error to think of dream, myth and art as being about any one matter other than relationship. As was mentioned earlier, dream is metaphoric and is not particularly about the relata mentioned in the dream. In the conventional interpretation of dream, another set of relata, often sexual, is substituted for the set in the dream. But perhaps by doing this we only create another dream. There indeed is no a priori reason for supposing that the sexual relata are any more primary or basic than any other set.

In general, artists are very unwilling to accept interpretations of this sort, and it is not clear that their objection is to the sexual nature of the interpretation. Rather, it seems that rigid focusing upon any single set of relata destroys for the artist the more profound significance of the work. If the picture were *only* about sex or *only* about social organization, it would be trivial. It is non-trivial or profound precisely because it is about sex and social organization and cremation, and other things. In a word, it is only about relationship and not about *any* identifiable relata. [...]

Gregory Bateson, extracts from 'Style, Grace and Information in Primitive Art', paper for the Wenner-Gren Conference on Primitive Art and Society (Burg Wartenstein, Austria, 1967); reprinted in Bateson, *Steps to an Ecology of Mind* (North Vale, New Jersey: Jason Aronson, Inc., 1972) 108–23; reprinted edition (Chicago: University of Chicago Press, 2000).

Mary Catherine Bateson
Our Own Metaphor//1972

[...] We had been moving in many ways towards a sense that a solution to the ecological crisis would demand a new definition of the individual, a definition by which the arguments of 'economic man' would cease to be relevant, one that would retain a sense of the vividness of persons and at the same time allow each person to identify with natural process. To learn to love, we would need to recognize ourselves as systems, the beloved as systemic, similar and lovely in complexity, and to see ourselves at the same time as merged in a single system with the beloved. [...]

'I want to say a sentence to you and then I want to interpret that sentence in [several] very different ways. [...] *Each person is his own central metaphor.*'

'The first thing that I want to mean by that has to do with perception and coding. Any kind of representation within a person of something outside depends on there being sufficient diversity within him to reflect the relationships in what he perceives, as it depends on coding of some kind. The possibility of seeing something, the possibility of talking about it, and probably the possibility of loving, depend in every case on arriving in yourself at a comparable complexity, which depends in turn on the kind of diversity existing within yourself. [...]

'Now, we've talked here about the fact that there are lots of different kinds of representations – that is, the relationships within this system that is me can be used to reflect or to map other relationships in a very large number of ways. [...] If we're going to talk about relationships instead of about things, then all our talk about what exists, what's prior to what, and so on, just has to be rethought completely. I mean this first interpretation of that statement of mine to relate to many of the things we've said about how errors get in, since the way in which something is coded determines the kind of errors you can get. We can't relate to anything unless we can express its complexity through the diversity that is ourselves.' We err through a mismatch between ourselves and the other, and all our falsehoods are falsehoods about ourselves as well.

'Okay, the second thing that I want to mean by my sentence', I went on, 'brings us to the edge of a lot of more anthropological and psychoanalytic ideas. See, most of the ways in which we mutilate the environment and muck up systems have to do with things that we dislike about ourselves. [...] Now I've tried to say that the whole possibility of our dealing with complexity at all has to do with the complexity that is within ourselves. We are extraordinarily beautiful, intricate beings, sets of relationships. If we could see ourselves in the intricacy that we are, not just the little bit that comes into consciousness, we would be, I think, worthy of worship, because that's the only way we can love or worship anything.

But if we include, about ourselves, all of the intricacy, all of the cycling, all of the being born and dying on various scales, right down to the processes [...] in our cells, we have a position from which to love other persons equally complicated, or an ecosystem, or anything else.

'I'd like just to refer here, without spelling it all out, to a whole lot of stuff about the relationship between body image and the way you see the world, the relationship between the rhythms in our bodies and the way we deal with the world.' [...]

'Now, the question of consciousness brings up the fact that we have incomplete access to the complexity that we are. We've blocked a great deal of it out. In a Freudian sense, we've blocked it out by rejecting it. We also – it eludes us, it's too fine-grained, we're just not organized to be aware of it. One reason why poetry is important for finding out about the world is because in poetry a set of relationships get mapped onto a level of diversity in us that we don't ordinarily have access to. We bring it out in poetry. We can give to each other in poetry the access to a set of relationships in the other person and the world that we're not usually conscious of in ourselves. So we need poetry as knowledge about the world and about ourselves, because of this mapping from complexity to complexity.'

'Daddy [Gregory Bateson], you talked about seeing an ecosystem as beautiful. Now, when we say we can see it as beautiful, that may be the only way that we can talk about the fact that we've perceived a set of relationships in it.'

'And in ourselves.'

'And in ourselves. That's the point. They have to be in ourselves to see them in it.' [...]

Mary Catherine Bateson, extracts from *Our Own Metaphor* (New York: Alfred A. Knopf, 1972; reprinted edition: Creskill, New Jersey: Hampton Press, 2004) 284–9.

Heinz von Foerster
Cybernetics of Cybernetics//1973

[...] Here is [a] proposition, which I shall now baptize 'Humberto Maturana's Theorem Number One':

'Anything said is said *by* an observer.'

Should you at first glance be unable to sense the profundity that hides behind the simplicity of this proposition let me remind you of [Charles] West Churchman's admonition of this afternoon: 'You will be surprised how much can be said by a tautology'. This, of course, he said in utter defiance of the logician's claim that a tautology says nothing.

I would like to add to Maturana's Theorem a corollary which, in all modesty, I shall call 'Heinz von Foerster's Corollary Number One':

'Anything said is said *to* an observer.'

With these two propositions a non-trivial connection between three concepts has been established. First, that of an *observer* who is characterized by being able to make descriptions. This is because of Theorem 1. Of course, what an observer says is a description. The second concept is that of *language*. Theorem 1 and Corollary 1 connect two observers through language. But, in turn, by this connection we have established the third concept I wish to consider, namely that of *society*: the two observers constitute the elementary nucleus for a society. Let me repeat the three concepts that are in a triadic fashion connected to each other. They are: first, the observers; second, the language they use; and third, the society they form by the use of their language. This interrelationship can be compared, perhaps, with the interrelationship between the chicken and the egg and the rooster. You cannot say who was first and you cannot say who was last. You need all three in order to have all three. In order to appreciate what I am going to say it might be advantageous to keep this closed triadic relation in mind.

I have no doubts that you share with me the conviction that the central problems of today are societal. On the other hand, the gigantic problem-solving conceptual apparatus that evolved in our Western culture is counter-productive not only for solving but essentially for perceiving social problems. One root for our cognitive blind spot that disables us to perceive social problems is the traditional explanatory paradigm which rests on two operations: One is *causation*, the other one *deduction*. It is interesting to note that something that cannot be explained – that is, for which we cannot show a cause or for which we do not have a reason – we do not wish to see. In other words, something that cannot be explained cannot be seen. This is driven home again and again by Don Juan, a Yaqui Indian, Carlos Casteneda's mentor. It is quite clear that in his teaching efforts Don Juan wants to make a cognitive blind spot in Castaneda's vision to be filled with new perceptions; he wants to make him 'see'. This is doubly difficult, because of Castaneda's dismissal of experiences as 'illusions' for which he has no explanations on the one hand, and because of a peculiar property of the logical structure of the phenomenon 'blind spot' on the other hand; and this is that we do not perceive our blind spot by, for instance, seeing a black spot close to the centre of our visual field: we do not see that we have a blind spot. In other words, we do not see that we do not see. This I will call a second order deficiency, and the only way to overcome such deficiencies is with therapies of second order.

The popularity of Carlos Castaneda's books suggest to me that his points are being understood: new paradigms emerge. I'm using the term 'paradigm' in the sense of Thomas Kuhn, who wants to indicate with this term a culture-specific, or language-specific, stereotype or model for linking descriptions semantically. As you may remember, Thomas Kuhn argues that there is a major change in paradigms when the one in vogue begins to fail, shows inconsistencies or contradictions. I, however, argue that I can name at least two instances in which not the emergent defectiveness of the dominant paradigm but its very flawlessness is the cause for its rejection. One of these instances was Copernicus' novel vision of a heliocentric planetary system which he perceived at a time when the Ptolemaic geocentric system was at its height as to accuracy of its predictions. The other instance, I submit, is being brought about today by some of us who cannot – by their life – pursue any longer the flawless but sterile path that explores the properties seen to reside within objects, and turn around to explore their very properties seen now to reside within the observer of these objects. Consider, for instance, 'obscenity'. There is at periodic intervals a ritual performed by the supreme judges of this land in which they attempt to establish once and for all a list of all the properties that define an obscene object or act. Since obscenity is not a property residing within things (for if we show Mr X a painting and he calls it obscene, we know a lot about Mr X but very little about the painting), when our lawmakers will finally come up with their imaginary list we shall know a lot about them but their laws will be dangerous nonsense.

With this I come now to the other root for our cognitive blind spot and this is a peculiar delusion within our Western tradition, namely, 'objectivity':

'The properties of the observer shall not enter the description of his observations.'

But I ask, how would it be possible to make a description in the first place if not the observer were to have properties that allows for a description to be made? Hence, I submit in all modesty, the claim for objectivity is nonsense! One might be tempted to negate 'objectivity' and stipulate now 'subjectivity'. But please remember that if a nonsensical proposition is negated, the result is again a nonsensical proposition. However, the nonsensicality of these propositions either in the affirmative or in their negation cannot be seen in the conceptual framework in which these propositions have been uttered. If this is the state of affairs, what can be done? We have to ask a new question:

'What are the properties of an observer?'

Let me at once draw your attention to the peculiar logic underlying this question. For whatever properties we may come up with it is we, you and I, who have to make this observation, that is, we have to observe our own observing, and ultimately account for our own accounting. Is this not opening the door for the logical mischief of propositions that refer to themselves ('I am a liar') that have been so successfully excluded by Bertrand Russell's Theory of Types not to bother us ever again? Yes and No!

It is most gratifying for me to report to you that the essential conceptual pillars for a theory of the observer have been worked out. The one is a calculus of infinite recursions;[1] the other one is a calculus of self-reference.[2] With these calculi we are now able to enter rigorously a conceptual framework which deals with *observing* and not only with the observed.

Earlier I proposed that a therapy of the second order has to be invented in order to deal with dysfunctions of the second order. I submit that the cybernetics of observed systems we may consider to be first-order cybernetics; while second-order cybernetics is the cybernetics of observing systems. This is in agreement with another formulation that has been given by Gordon Pask. He, too, distinguishes two orders of analysis. The one in which the observer enters the system by stipulating the *system*'s purpose. We may call this a 'first-order stipulation'. In a 'second-order stipulation' the observer enters the system by stipulating *his own* purpose.

From this it appears to be clear that social cybernetics must be a second-order cybernetics – a *cybernetics of cybernetics* – in order that the observer who enters the system shall be allowed to stipulate his own purpose: he is autonomous. If we fail to do so somebody else will determine a purpose for us. Moreover, if we fail to do so, we shall provide the excuses for those who want to transfer the responsibility for their own actions to somebody else: 'I am not responsible for my actions; I just obey orders.' Finally, if we fail to recognize autonomy of each, we may turn into a society that attempts to honour commitments and forgets about its responsibilities. [...]

1 [reference 11 in source] Paul E. Weston and Heinz von Foerster, 'Artificial Intelligence and Machines That Understand', in H. Eyring, C.H. Christensen and H.S. Johnston, eds, *Annual Review of Physical Chemistry*, no. 24 (Palo Alto: Annual Review Inc., 1973) 358–78.

2 [9] Francisco Varela, 'A Calculus for Self-reference', *International Journal of General Systems*, vol. 2, no. 1 (1975) 1–25.

Heinz von Foerster, extract from abbreviated version of 'Cybernetics of Cybernetics', paper for fourth International Conference on Design Research (Blacksburg: Virginia Polytechnic Institute, 15 April 1973), in *Communication and Control*, ed. Klaus Krippendorff (New York: Gordon and Breech, 1979) 5–8.

Humberto Maturana and Francisco Varela
The Tree of Knowledge: Biological Roots of Human Understanding//1984

The act of indicating any being, object, thing or unity involves making *an act of distinction* which distinguishes what has been indicated as separate from its background. Each time we refer to anything explicitly or implicitly, we are specifying a *criterion of distinction*, which indicates what we are talking about and specifies its properties as being, unity or object.

This is a commonplace situation and not unique: we are necessarily and permanently immersed in it.

A unity (entity, object) is brought forth by an act of distinction. Conversely, each time we refer to a unity in our descriptions, we are implying the operation of distinction that defines it and makes it possible. [...]

When we speak of living beings, we presuppose something in common between them; otherwise we wouldn't put them in the same class we designate with the name 'living'. What has not been said, however, is: What is that organization that defines them as a class? Our proposition is that living beings are characterized in that, literally, they are continually self-producing. We indicate this process when we call the organization that defines them an *autopoietic organization*. [...]

The most striking feature of an autopoietic system is that it pulls itself up by its own bootstraps and becomes distinct from its environment through its own dynamics, in such a way that both things are inseparable.

Living beings are characterized by their autopoietic organization. They differ from each other in their structure, but they are alike in their organization.

By realizing what characterizes living beings in their autopoietic organization, we can unify a whole lot of empirical data about [them]. The concept of autopoiesis [...] explicitly proposes that such data be interpreted from a specific point of view which stresses that living beings are *autonomous* unities.

We use the word 'autonomy' in its current sense; that is, a system is autonomous if it can specify its own laws, what is proper to it. We are *not* proposing that living beings are the only autonomous entities. Certainly they are not. But one of the most evident features of a living being is its autonomy. We *are* proposing that the mechanism that makes living beings autonomous systems is autopoiesis. This characterizes them as autonomous systems. [...]

That living beings have an organization, of course, is proper not only to them but also to everything we can analyse as a system. What is distinctive about them, however, is that their organization is such that their only product is themselves, with no separation between producer and product. The being and doing of an autopoietic unity are inseparable, and this is their specific mode of organization. [...]

Ontogeny is the history of structural change in a unity without loss of organization in that unity. This ongoing structural change occurs in the unity from moment to moment, either as a change triggered by interactions coming from the environment in which it exists or as a result of its internal dynamics. As regards its continuous interactions with the environment, the cell unity classifies them and sees them in accordance with its structure at every instant. That structure, in turn, continuously changes because of its internal dynamics. The overall result is that the ontogenic transformation of a unity ceases only with its disintegration. [...]

Now, what happens when we consider the ontogeny of, not one, but two (or more) neighbouring unities in their medium of interaction?

We can look at this situation, of course, from the perspective of either one of the unities, and it will be symmetrical. This means that, for the cell on the left, the one on the right is only one more source of interactions, indistinguishable from those which we, as observers, classify as coming from the 'inert' environment. Conversely, for the cell on the right, the other is one more source of interactions encountered according to its own structure.

This means that two (or more) autopoietic unities can undergo coupled ontogenies when their interactions take on a recurrent or more stable nature. We have to keep this clearly in mind. Every ontogeny occurs within an environment; we, as observers, can describe both as having a particular structure such as diffusion, secretion, temperature. In describing autopoietic unity as having a particular structure, it will become clear to us that the interactions (as long as they are recurrent) between unity and environment will consist of reciprocal perturbations. In these interactions, the structure of the environment only triggers structural changes in the autopoietic unities (it does not specify or direct them), and vice versa for the environment. The result will be a history of mutual congruent structural changes as long as the autopoietic unity and its containing environment do not disintegrate: there will be a structural coupling. [...]

Our discussion has led us to conclude that, biologically, there is no 'transmitted information' in communication. Communication takes place each time there is behavioural coordination in a realm of structural coupling.

This conclusion is surprising only if we insist on not questioning the latest metaphor for communication which has become popular with the so-called communication media. According to this metaphor of the tube, communication is something generated at a certain point. It is carried by a conduit (or tube) and is delivered to the receiver at the other end. Hence there is a *something* that is communicated, and what is communicated is an integral part of that which travels in the tube. Thus we usually speak of the 'information' contained in a picture, an object, or, more evidently, the printed word.

According to our analysis, this metaphor is basically false. It presupposes a unity that is not determined structurally, where interactions are instructive, as though what happens to a system in an interaction is determined by the perturbing agent and not by its structural dynamics. It is evident, however, even in daily life, that such is not the case with communication: each person says what he says or hears what he hears according to his own structural determination; saying does not ensure listening. From the perspective of an observer, there is always ambiguity in a communicative interaction. The phenomenon of communication depends on not what is transmitted, but on what happens to the person who receives it. And this is a very different matter from 'transmitting information'. [...]

Organisms and societies belong to one class of metasystems; these consist of aggregates of autonomous unities that can be cellular or metacellular. An observer can distinguish the different metasystems of this class by the different degrees of autonomy he sees possible in their components. Thus, if he should put them in a series according to the degree of dependency of their components (in their embodiment as autonomous unities) on their participation in the metasystem they form, organisms and human social systems would be at the opposite ends of the series. Organisms would be metasystems of components with minimum autonomy, i.e., components with very little or no dimension of independent existence. [...]

As metacellular systems, organisms have operational closure in the reciprocal structural coupling of their component cells. The central feature in the organization of an organism lies in its manner of being a unity in an environment wherein it must operate with stable properties that permit it to conserve its adaptation, whatever the properties of its components may be. [...]

[H]uman social systems [...] have operational closure, too, in the structural coupling of their components. But human social systems exist also as unities for their components in the realm of language. Therefore, the identity of human social systems depends on the conservation of adaptation of human beings not only as organisms (in a general sense) but also as components of their linguistic domains. [...]

[C]entral to the operation of a human social system is the linguistic domain that its components generate and the extension of their properties – a condition necessary for the embodiment of language, which is their realm or domain of existence. The organism restricts the individual creativity of its component unities, as these unities exist for that organism. The human social system amplifies the individual creativity of its components, as that system exists for these components. [...]

Coherence and harmony in relations and interactions between the members of a human social system are due to the coherence and harmony of their growth in it, in an ongoing social learning which their own social (linguistic) operation defines and which is possible thanks to the genetic and ontogenetic processes that permit structural plasticity of the members. [...]

Humberto Maturana and Francisco Varela, extracts from *The Tree of Knowledge: Biological Roots of Human Understanding* (Boston: New Science Library, 1984); reprinted edition (Boston: Shambhala Publications, Inc., 1987) 40–48, 74–5, 198–9.

Donella H. Meadows
Dancing with Systems//2001

[...] People who are raised in the industrial world and who get enthused about systems thinking are likely to make a terrible mistake. They are likely to assume that here, in systems analysis, in interconnection and complication, in the power of the computer, here at last, is the key to prediction and control. This mistake is likely because the mindset of the industrial world assumes that there is a key to prediction and control. [...]

But self-organizing, non-linear, feedback systems are inherently unpredictable. They are not controllable. They are understandable only in the most general way. The goal of foreseeing the future exactly and preparing for it perfectly is unrealizable. The idea of making a complex system do just what you want it to do can be achieved only temporarily, at best. We can never fully understand our world, not in the way our reductionistic science has led us to expect. Our science itself, from quantum theory to the mathematics of chaos, leads us into irreducible uncertainty. For any objective other than the most trivial, we can't optimize; we don't even know what to optimize. We can't keep track of everything. We can't find a proper, sustainable relationship to nature, each other, or the institutions we create, if we try to do it from the role of omniscient conqueror. [...]

Systems thinking leads to another conclusion, however – waiting, shining, obvious, as soon as we stop being blinded by the illusion of control. It says that there is plenty to do, of a different sort of 'doing'. The future can't be predicted, but it can be envisioned and brought lovingly into being.

Systems can't be controlled, but they can be designed and redesigned. We can't surge forward with certainty into a world of no surprises, but we can expect surprises and learn from them and even profit from them. We can't impose our will upon a system. We can listen to what the system tells us, and discover how its properties and our values can work together to bring forth something much better than could ever be produced by our will alone.

We can't control systems or figure them out. But we can dance with them!

I already knew that, in a way before I began to study systems. I had learned about dancing with great powers from whitewater kayaking, from gardening, from playing music, from skiing. All those endeavours require one to stay wide-awake, pay close attention, participate flat-out and respond to feedback. It had never occurred to me that those same requirements might apply to intellectual work, to management, to government, to getting along with people.

But there it was, the message emerging from every computer model we made.

Living successfully in a world of systems requires more of us than our ability to calculate. It requires our full humanity – our rationality, our ability to sort out truth from falsehood, our intuition, our compassion, our vision and our morality. [...]

1. Get the beat

Before you disturb the system in any way, watch how it behaves. If it's a piece of music or a whitewater rapid or a fluctuation in a commodity price, study its beat. If it's a social system, watch it work. Learn its history. Ask people who've been around a long time to tell you what has happened. [...]

Starting with the behaviour of the system forces you to focus on facts, not theories. It keeps you from falling too quickly into your own beliefs or misconceptions, or those of others. [...]

Starting with the behaviour of the system directs one's thoughts to dynamic, not static analysis – not only to 'what's wrong?' but also to 'how did we get there?' and 'what behaviour modes are possible?' and 'if we don't change direction, where are we going to end up?' [...]

2. Listen to the wisdom of the system

Aid and encourage the forces and structures that help the system run itself. Don't be an unthinking intervener and destroy the system's own self-maintenance capacities. Before you charge in to make things better, pay attention to the value of what's already there. [...]

3. Expose your mental models to the open air

Remember, always, that everything you know, and everything everyone knows, is only a model. Get your model out there where it can be shot at. Invite others to challenge your assumptions and add their own. Instead of becoming a champion for one possible explanation or hypothesis or model, collect as many as possible. Consider all of them plausible until you find some evidence that causes you to rule one out. That way you will be emotionally able to see the evidence that rules out an assumption with which you might have confused your own identity. [...]

4. Stay humble. Stay a learner

Systems thinking has taught me to trust my intuition more and my figuring-out rationality less, to lean on both as much as I can, but still to be prepared for surprises. Working with systems, on the computer, in nature, among people, in organizations, constantly reminds me of how incomplete my mental models are, how complex the world is, and how much I don't know.

The thing to do, when you don't know, is not to bluff and not to freeze, but to learn. The way you learn is by experiment – or, as Buckminster Fuller put it, by trial and error, error, error. In a world of complex systems it is not appropriate to charge forward with rigid, undeviating directives. 'Stay the course' is only a good idea if you're sure you're on course. Pretending you're in control even when you aren't is a recipe not only for mistakes, but for not learning from mistakes. What's appropriate when you're learning is small steps, constant monitoring, and a willingness to change course as you find out more about where it's leading. [...]

5. Honour and protect information

A decision-maker can't respond to information he or she doesn't have, can't respond accurately to information that is inaccurate, can't respond in a timely way to information that is late. I would guess that 99 percent of what goes wrong in systems goes wrong because of faulty or missing information.

If I could, I would add an Eleventh Commandment: Thou shalt not distort, delay or sequester information. You can drive a system crazy by muddying its information streams. You can make a system work better with surprising ease if you can give it more timely, more accurate, more complete information. [...]

6. Locate responsibility in the system

Look for the ways the system creates its own behaviour. Do pay attention to the triggering events, the outside influences that bring forth one kind of behaviour from the system rather than another. Sometimes those outside events can be controlled (as in reducing the pathogens in drinking water to keep down incidences of infectious disease). But sometimes they can't. And sometimes blaming or trying to control the outside influence blinds one to the easier task of increasing responsibility within the system.

'Intrinsic responsibility' means that the system is designed to send feedback about the consequences of decision-making directly and quickly and compellingly to the decision-makers. [...]

7. Make feedback policies for feedback systems

[...] You can imagine why a dynamic, self-adjusting system cannot be governed by a static, unbending policy. It's easier, more effective, and usually much cheaper to design policies that change depending on the state of the system. Especially where there are great uncertainties, the best policies not only contain feedback loops, but meta-feedback loops – loops that alter, correct and expand loops. These are policies that design learning into the management process.

8. Pay attention to what is important, not just what is quantifiable

Our culture, obsessed with numbers, has given us the idea that what we can measure is more important than what we can't measure. You can look around and make up your own mind about whether quantity or quality is the outstanding characteristic of the world in which you live.

If something is ugly, say so. If it is tacky, inappropriate, out of proportion, unsustainable, morally degrading, ecologically impoverishing, or humanly demeaning, don't let it pass. Don't be stopped by the 'if you can't define it and measure it, I don't have to pay attention to it' ploy. No one can precisely define or measure justice, democracy, security, freedom, truth or love. No one can precisely define or measure any value. But if no one speaks up for them, if systems aren't designed to produce them, if we don't speak about them and point toward their presence or absence, they will cease to exist.

9. Go for the good of the whole

Don't maximize parts of systems or subsystems while ignoring the whole. [...] Aim to enhance total systems properties, such as creativity, stability, diversity, resilience and sustainability – whether they are easily measured or not.

As you think about a system, spend part of your time from a vantage point that lets you see the whole system, not just the problem that may have drawn you to focus on the system to begin with. And realize that, especially in the short term, changes for the good of the whole may sometimes seem to be counter to the interests of a part of the system. It helps to remember that the parts of a system cannot survive without the whole. [...]

10. Expand time horizons

The official time horizon of industrial society doesn't extend beyond what will happen after the next election or beyond the payback period of current investments. The time horizon of most families still extends farther than that – through the lifetimes of children or grandchildren. [...] The longer the operant time horizon, the better the chances for survival.

In the strict systems sense there is no long-term/short-term distinction. Phenomena at different timescales are nested within each other. Actions taken now have some immediate effects and some that radiate out for decades to come. We experience now the consequences of actions set in motion yesterday and decades ago and centuries ago. [...]

11. Expand thought horizons

[...] Seeing systems whole requires more than being 'interdisciplinary', if that word means, as it usually does, putting together people from different disciplines and letting them talk past each other. Interdisciplinary communication works only if there is a real problem to be solved, and if the representatives from the various disciplines are more committed to solving the problem than to being academically correct. They will have to go into learning mode, to admit ignorance and be willing to be taught, by each other and by the system.

It can be done. It's very exciting when it happens.

12. Expand the boundary of caring

Living successfully in a world of complex systems means expanding not only time horizons and thought horizons; above all it means expanding the horizons of caring. There are moral reasons for doing that, of course. And if moral arguments are not sufficient, then systems thinking provides the practical reasons to back up the moral ones. The real system is interconnected. No part of the human race is separate either from other human beings or from the global ecosystem. It will not be possible in this integrated world for your heart to succeed if your lungs fail, or for your company to succeed if your workers fail, or for the rich in Los Angeles to succeed if the poor in Los Angeles fail, or for Europe to succeed if Africa fails, or for the global economy to succeed if the global environment fails. [...]

13. Celebrate complexity

[...] Let's face it, the universe is messy. It is non-linear, turbulent and chaotic. It is dynamic. It spends its time in transient behaviour on its way to somewhere else, not in mathematically neat equilibria. It self-organizes and evolves. It creates diversity, not uniformity. That's what makes the world interesting, that's what makes it beautiful, and that's what makes it work. [...]

14. Hold fast to the goal of goodness

[...] The gap between desired behaviour and actual behaviour narrows. Fewer actions are taken to affirm and instill ideals. The public discourse is full of cynicism. Public leaders are visibly, unrepentantly, amoral or immoral and are not held to account. Idealism is ridiculed. Statements of moral belief are suspect. It is much easier to talk about hate in public than to talk about love.

We know what to do about eroding goals. Don't weigh the bad news more heavily than the good. And keep standards absolute. [...]

And so we are brought to the gap between understanding and implementation. Systems thinking by itself cannot bridge that gap. But it can lead us to the edge of what analysis can do and then point beyond – to what can and must be done by the human spirit.

Donella H. Meadows, extracts from 'Dancing with Systems' (text excerpted from the manuscript of the author's last, unfinished book).

CYBERNATED

ART IS VERY IMPORTANT
BUT ART FOR

CYBERNATED

LIFE IS MORE IMPORTANT
AND THE LATTER NEED NOT BE

CYBERNATED

CYBERNETIC ART, ARCHITECTURE AND DESIGN

Nam June Paik
Cybernated Art//1966

Cybernated art is very important, but art for cybernated life is more important, and the latter need not be cybernated.

(Maybe George Brecht's *simplissimo* is the most adequate.)

But if Pasteur and Robespierre are right that we can resist poison only through certain built-in poison, then some specific frustrations, caused by cybernated life, require accordingly cybernated shock and catharsis. My everyday work with video tape and the cathode-ray tube convinces me of this.

Cybernetics, the science of pure relations, or relationship itself, has its origin in karma. Marshall McLuhan's famous phrase 'Media is message' was formulated by Norbert Wiener in 1948 as 'The signal, where the message is sent, plays equally important role as the signal, where the message is not sent.'

As the Happening is the fusion of various arts, so cybernetics is the exploitation of boundary regions between and across various existing sciences.

Newton's physics is the mechanics of power and the unconciliatory two-party system, in which the strong win over the weak. But in the 1920s a German genius put a tiny third-party (grid) between these two mighty poles (cathode and anode) in a vacuum tube, thus enabling the weak to win over the strong for the first time in human history. It might be a Buddhistic 'third way', but anyway this German invention led to cybernetics, which came to the world in the last war to shoot down German planes from the English sky.

The Buddhists also say
Karma is samsara
Relationship is metempsychosis

We are in open circuits

Nam June Paik, 'Cybernated Art', in *Manifestos* (Great Bear Pamphlets series) (New York: Something Else Press, 1966) 24; reprinted in Kristine Stiles and Peter Selz, eds, *Theories and Documents of Contemporary Art* (Berkeley and Los Angeles: University of California Press, 1996) 433–4 [in the original, symbols accompany the text].

Roy Ascott
The Cybernetic Stance: My Process and Purpose//1968

[…] *As distinctions between music, painting, poetry, etc. become blurred and media are mixed, a* behaviourist synthesis *is seen to evolve, in which dialogue and feedback within a social culture indicate the emergence of a Cybernetic vision in art as in science.* […]

The Cybernetic Art Matrix is seen as a process in which anarchic group behaviour interacts with pre-established systems of communications, hardware and learning nets. In both cases the processes are self-generating and self-critical. Basically they are initiated by creative behaviour, and in turn give rise to its extension in other people.

I

The paradox we face as artists writing about our work is that the future is all that interests us, and that is precisely the part of our activity which must remain necessarily unpredictable. For many of us there is a further paradox; we can see that, as the value of the commercial gallery declines and our interest in the art object as *object* diminishes, so the need for new channels of communication between people increases, free access to new technology and media becomes imperative, and a new cultural situation inexorably evolves. The paradox lies in the fact that we continue to throw off art objects in the course of our creative work, while our eyes are set on the new horizon. We have undoubtedly become *process-oriented* but we still deal with objects. […]

I am concerned, in short, with some kind of tangible philosophy, with ideas in action. Both CAM [Cybernetic Art Matrix] on a social scale and my individual artefacts on a intimate scale are essentially *triggers*. They contain nothing but the possibility of future action; that is to say they exist only in so far as the spectator participates in their evolution by, on the one hand, interacting with other people within a complex social situation, and on the other hand by conducting a private interior dialogue.

If the modern era in science and art and human affairs can be differentiated from previous eras, as the outward aspect of events would suggest, it must contain some unifying quality, some basic characteristics which are shared by artists, scientists and politicians alike. If we examine the apparently diverse tendencies in all these fields, we may discern a common vision. This vision, still clouded and imprecise, is characterized by a fundamental and mutual tendency: a tendency to the creation of *dialogue*. In previous periods of western society, art, science and politics have tended to be deterministic, absolutist, hierarchic. The channels of communication have been one-way channels, the flow of information has been in one direction. In each area of activity we find a closed system: an image is projected, a principle expounded, a social relationship established, in each case, only to reinforce a fixed point of view, an absolute ideal, a permanent set of values.

But now *change* is everywhere apparent. Human beings are mobile geographically and socially; the scientist not only observes an experiment, he participates in it; the artist's interest lies more in the process of working than in the finished art work, and his audience expects, not a fixed attitude or viewpoint to the work, but a field of uncertainty and ambiguity in which they can, endlessly, take part. In every area the system, so regarded, is open-ended; nothing is fixed. Today we are concerned less with the essence of things as with their behaviour; not what they are but what they do. This unified tendency is evidently *behavioural*, and we can see how the vision of our time is ultimately cybernetic.

This new vision contrasts forcibly with the old attitudes. Henri Bergson [...] revealed repeatedly in his *Philosophy of Change* the nature of the new situation. 'The role of life is to insert some *indetermination* into matter' ... 'The living are relatively stable, and counterfeit immobility so well that we treat each of them as thing rather than as progress, forgetting that the very permanence of their form is only the outline of a movement.'[1] A shift of human interest is from the thing, the object, the product, to the process, the system, the event in which the product is obtained. The history of modern art, with its roots in that same period, is the history of this shift; a radical move which, as it evolves, may take us into a culture more exhilarating and free than previously we might have imagined possible.

II

I am suggesting that a *behaviourist* framework can be constructed from which to examine, not only the internal relations of modern art, but its social implications, and its potential contribution to our forming cybernetic culture. Everywhere in modern art, particularly in the visual/plastic arts, but also in the more experimental reaches of music and literature, the emphasis is on behaviour, on what happens, on process and system, the dynamic interplay of random and ordered elements. [...]

As for the spectator, he no longer expects to receive a ready-made experience, or the expression of an experience, but rather to participate at a deep level, either in his consciousness or, more physically, by immediate action. The artist no longer decides everything and projects it as a whole in some definitive and final composition; he now initiates a dialogue, or set of events, which, when taken up by the audience, whether in a group or individually, will be shaped into totally unpredictable and indeterminate forms and experiences. [...]

We are entering an era in which everyone takes responsibility for the common culture, by participating in the decisions and actions which will inform it.

As feedback between persons increases and communications become more rapid and precise, so the creative process no longer culminates in the *artwork*, but extends beyond it deep into the life of each individual. Art is then determined not by the creativity of the artist alone, but by the creative behaviour his work induces in the spectator, and in society at large. Where art of the old order constituted a *deterministic vision*, so the art of our time tends towards the development of a cybernetic vision, in which feedback, dialogue and involvement in some creative interplay at deep levels of experience are paramount. But a *vision* of our time, if it is truly representative, must embrace more than the aspirations of its artists; it will include scientists, technologists, economists, all those fields of human endeavour in which creativity is prime. And a common spirit which can be called *cybernetic* infuses all these fields today. [...]

[T]he cybernetic spirit, more than the method or the applied science, creates a continuum of experience and knowledge which radically reshapes our philosophy, influences our behaviour and extends our thought. [...]

There can be no doubt that the *cybernetic vision*, as it emerges in our consciousness, will rapidly effect great changes in the human condition.

III

I have suggested that modern art may be best understood if it is examined in the context of behaviour, that there is a forming aesthetic of *process* and *system* in which the cardinal factors of feedback and interplay are consistent with a *cybernetic vision*. [...]

Where the behaviour of the artist is uppermost, where the focus is on the artist's activity for its own sake, as for example in the case of Jackson Pollock, the action painter in his arena, we can see that aspect of art as *behavioural ritual*. Action in which chance plays a large part was a characteristic of Surrealism, and, as in the case of Duchamp, the act of random choice can become ritualized to the degree of dispensing with the fabrication of art objects altogether. Once the action or event becomes all-important, 'happenings' are an inevitable consequence. People interacting freely in groups, producing unfamiliar situations, catalysing perhaps further group responses, constitute an art situation.

Where the artist is interested less in his own behaviour than in the behaviour of the spectator his work may be seen specifically as a *behavioural trigger*. The response in an observer might be elicited in a number of ways: *physically*, in the immediate sense of a highly intensive optical activity induced by visual stimuli creating flicker, after-images, spatial ambiguity and uncertainty of figure-ground relationships; or *manually*, where the nature of the artefact induces the spectator to alter the position of its parts; or again a *postural response* may be effected, with the observer moving about, shifting position, so that images merge and transform, and in so doing encourage this activity as the sole relationship possible between artefact and the human being.

Artworks may also trigger off responses of a polemical or social kind, encouraging in their audience changes in individual or group behaviour by questioning preconceptions, destroying illusions by means of the shock of unfamiliar, absurd or incongruous imagery. Within the same context a more sober, dialectical construction of abstract values has been developed to present the observer with the possibility of new social behaviour.

Again, the artist's main interest may lie not in his own behaviour, nor in that of the observer, but more especially in the behaviour of the objects he makes. These *behavioural structures* literally behave in the sense of articulating their various parts in response to internal or external stimuli. The medium may be light moving onto surfaces, themselves moving perhaps, or fins of metal dependent on the impact of air currents to push themselves round. They may be structures internally powered by electricity or some hydraulic device. Another possibility, dramatically demonstrated by Tinguely [i.e. *Homage to New York*, 1960], is the built-in capacity systematically to destroy itself. We must look to the future and the research of, for example, Stafford Beer or Gordon Pask, for those 'fungoid systems' and chemical-colloidal computors which might make possible the creation of *behavioural structures* invested with the properties of growth. [...]

The *behaviourist tendency* [...] implies a *total* behavioural involvement in which all our senses are brought into play, not simply visual, but postural, tactile and including the sense of hearing and even of taste and smell. In short, a *behaviourist synthesis* is forming where the boundaries between the once highly differentiated arts of music, poetry, painting, architecture, sculpture and acting are becoming less distinct. The media merge, and at the same time the distinction between the roles of artist and audience becomes blurred. The artwork or event is a matrix between two sets of behaviour, which through it become one, continuous and interrelated. Inevitably a state of perfect feedback will emerge, where we all both initiate and involve ourselves in total creative situations. [...]

IV

I have already said that both these artefacts and the CAM-complex are intended to function as *triggers*, and that the former are in some good sense models for the latter. They are part of the same overall process, they proceed from the same *stance*. That is, to initiate dialogue, to involve other people in creative behaviour, engaging more of the senses than the purely visual one alone. Presenting, not a set of ideas or a personal expression of feelings, but a situation in which other people's ideas and feelings can be set in motion, generating quite unpredictable experience. This process is deterministic only in the sense that it is aimed at eliciting creative thought and action in the community. The specifics of that action and of the directions my processes may take are equally unpredictable and open-ended. Each situation, as it emerges, will have its own controlling energy, and thus my work, in its implications, is essentially cybernetic. A function of the output variable (social, individual response) is to act as an input variable in my working process and in the art-work/experience, so introducing continually more and more variety into the system. [...]

1 Henri Bergson, *Creative Evolution* (1907), trans. A. Mitchell (London: Macmillan, 1920) 132–5. [*The Philosophy of Change* is the title of a study of Bergson by H.W. Carr, published in London in 1912.]

Roy Ascott, extracts from 'The Cybernetic Stance: My Process and Purpose', *Leonardo*, vol. 1, no. 2 (April 1968) 105–12.

Stephen Willats
Art Society Feedback: In Conversation with Emily Pethick//2011

Emily Pethick [...] You were one of the first British artists to take work out of galleries into society, making the audience active and examining the social functions and meanings of art in society. The relationship between practice and theory, especially in the diagrams, is a constant in your work since the late 1950s. One early piece, *Art Society Feedback* (1959), shows a conceptual model of connections between the artist and the social context, the feedback between artist and environment. This is an example of the conceptual models you create, which feed into works made in collaboration with people in social contexts outside the gallery.

Stephen Willats My development of concepts and models is related to a perception of the function of the artist in relation to the world. I was conscious of the fact that the work I was proposing was going to be part of society. An initial observation was that artwork is completely dependent on its audience. We could almost say viewers are its reason for being; without them it doesn't exist. It is essential for artists to realize they are somehow part of society. The next observation was that most art practice was describing existing values and beliefs, amplifying what was validated in existing society. Then there was another smaller, much more difficult but ultimately more meaningful role, concerning transformation; the notion that the artist can transform existing values and provide a vision of the future, a different perception of the world, and a language for that. I saw that practice was nothing more than a vehicle, embodying the language. You have got to have a model to represent reality; models are representations of an external, encountered or possible reality.

Pethick Diagrams have been a central aspect of your practice from very early on. They are used as a language for forming models, but also as a tool for planning projects. Your use of them was also influenced by exposure to theories outside art, like cybernetics, systems theories, black box theory.

Willats The diagram is a dynamic picture, a model in a dynamic state. I saw that other languages were needed to provide a vision of a future possible world. The languages available to me in the world of historical art were inadequate to describe the new reality, the new world I was encountering (in the late 1950s and early 60s), that seemed to be emerging. So I became interested in languages from outside art. The emerging sciences of cybernetics and information theory were especially exciting, as were the nascent philosophies of semiotics. All kinds of new ways of thinking were appearing and could be drawn into practice. It was just a natural way of representing ideas and social relationships in a dynamic way. If the artist was in a relationship with the audience, and the audience was part of society, the artist was in a relationship with society, so there was feedback. This is how my diagrams originated.

Pethick One of the most striking early works is *Homeostat Drawing* (1969). This diagram depicts an endless network of interconnecting parts. Can you talk about where this came from, what it represents as a social model?

Willats In the mid 1960s I encountered the work of Ross Ashby, who developed the homeostat. His representation was a model with four nodes, totally interconnected by input-output relationships. The important thing for me was that it showed a possible form of relationships and information within society. Though I don't think Ashby saw it this way; his was a mechanical model. Nevertheless I could glimpse social ramifications. The homeostat model posed another notion, illustrating the difference between our historical systems of control – where information is contained within a set hierarchy – and the idea of a continually shifting, self-determining system. This is another model of control, to make information available throughout structures, so this one-layer network could be seen as a new social model. I was interested in the notion of another society, moving away from the straitjacket I perceived in the 1950s. This early work led to simulation works showing a decision-making model of society based on mutual cooperation, like *Visual Homeostatic Information Mesh* (1969) and *Visual Homeostatic Maze* (1968). These simulations represented the self-organizing model of society in a dynamic state, and involved people in making decisions about their relationships with others. I have always been interested in cooperation, a comparative critique between competition and cooperation in decision making.

Pethick The *Homeostat Drawing* is also based on an idea of agreement, a frequent notion in your work, especially in later works like *Meta Filter* (1973–75).

Willats Yes, I saw agreement as a fundamental state. Agreement is not compliance, acquiescence; it involves perceptual recognition of mutuality. It requires a complex series of exchanges. Agreement is a social state between people, not a mechanistic thing; if one is conforming it may seem like agreement, but it is not.

Pethick The concept of self-organization is something you were interested in very early on, and explored in different ways: the individual's capacity to self-organize; non-conformity regarding imposed social structures. Resistance to control emerges when you look, for example, at the planned environment, at how tower blocks or modernist housing structure people's lives, how their inhabitants develop their own subcultures and languages.

Willats Absolutely. In the early 1970s I was consciously looking for polemics to represent in my work, and I thought about externalizing these observations and ideas. I saw that people were in a state of what I call counter-consciousness: they lived in a reality that was determined for them in a mechanistic way. They had to adapt to it, so they created their own counterculture. I don't think this movement or force was rationalized, that came later with the post-punks in the early 80s; it was a sort of basic human reaction to a crushing state of determinism. I felt the spirit of self-organization was alive even in the most depressing environments. I noticed this in tower blocks, where residents were isolated from external reality, physically and socially, but still fought back and managed to create a kind of symbolic society for themselves, to find mutual relationships. The development of a counter-consciousness was really important; without it, people would have collapsed. It helped them to maintain their own identity.

Pethick You have talked a lot about the artist as someone concerned with transformation, also in relation to the individual's capacity to transform, through the works you have made on housing estates.

Willats I saw my practice as a way of engaging with other people, forwarding a vision of society that has to be in a language people can understand. The traditional art world has its own special, exclusive languages. People know this, and I wanted to extend the meaning of my work beyond this exclusive environment. I had to find a way to build a bridge and make my propositions meaningful to them. That meant creating a symbolic world for an audience to enter, and articulating this symbolic world in a familiar language. The most appropriate language was their language, so embodying the audience's language in the work helped me to create the symbolic world too. One thing was to set up a relationship with the audience, a feedback between creator and observer of the work. The audience entering the symbolic world could make inferences to their own reality, looking at the world around them, then seeing how it could be transformed. I was working with people on the margins, alienated from the normal, predetermined behaviour of society. At the time, people said this was crazy and tried to stop me, but I said no: these people are important to the future because they embody the act of transformation, developing other languages to denote other ways of viewing future society. [...]

Pethick Reading your collected writings in the book *Art Society Feedback* (2010), there are some recurring principles of your work that become clear, connected to a reluctance to see things as fixed, or from one perspective, favouring dynamic states, open systems, acknowledging the complexity of people and experiences. This is something that can be addressed using the language of the diagram, but also through working with people, involving multiple authors. To explore more than one perspective, coexistences, multiple channels, uncertainties. Something very striking in your work is that you have often resisted a singular, authoritative perspective, in favour of open situations.

Willats What is fundamental to these models is the idea of self-organization and cooperation. I'm interested in acknowledgement of relativity, transience, fluidity, complexity. I think in the last three decades some very important things have become guiding principles. These ideas didn't exist in the 1930s, 40s or 50s. Last-century thinking said the world was simple, authoritative, monumental, immortal, etc., but in the world currently opening up before us we acknowledge the richness of complexity, transience, multi-channel fluidity, self-organization.

Stephen Willats and Emily Pethick, extracts from 'Art Society Feedback', interview, *Mousse Magazine*, no. 27 (January 2011). (http://moussemagazine.it)

Steina and Woody Vasulka

Woody's Famous Feedback Rap//1973

Woody [...] We look at video feedback as electronic art material. It's a building material for an image. It's totally abundant in its electronic nature. It's the clay, it's the air, it's the energy, it's the stone, it's the raw material that you simply use, and then build an image with it. And video feedback is very much what audio feedback is about. You use the relationship between a camera and a monitor the same way you use the relationship between a speaker and a microphone. An ambient noise is amplified through this cycle, and gets amplified further and further, until it results in almost unbearable sound. In video, of course, it happens on a much smaller volume scale. The image basically builds, it increases its volume, but at a certain point it does not increase anymore. It discharges its cycle, and then builds the image again and again.

Now you can of course influence the process: the speed of development or the direction of the feedback within a field; you can influence the left to right or the right to left development, in the sense of a spiral. These are all things, through discovery of working with feedback and through sets of errors, you can define what controls of the camera/monitor relationship to use to shape the feedback. Now we usually work with brightness, with f/stop on the lens, and the zoom: these are the major things, and then with the position of the monitor to the camera, or the position of the camera to the monitor. We can increase the intensity of the development and its complexity. And we can also simplify. We can somehow filter certain details out of the process. Feedback does not always have to develop into organic flow, like clay or electronic matter which increases in strange blobs of light. It can also be used simply as a mirror effect, if you zoom in and out you can see the frame; you can use the feedback as a mirror effect which again has its directions and laws. If you turn the camera, you get a curving effect, you can go up and down, and you can actually control that.

Now, if you use multi-monitors, or split screen effects, you can of course influence with different images the final composition. What's interesting about feedback is that any part of the image changes the composition of the whole frame. If a person walks in feedback, every minor movement or position within that frame organically changes the whole structure, even if it's not detectable. So that means the integration between the object and electronic feedback is total, there's no division. It's not a passive process, superimposition, or parts of collaging or matting. It's an organic influence to the image. Now, we usually don't work with a single feedback. We usually use feedback as part of the frame, or, as in the last composition we did, the feedback as a set, controlled in a monitor becoming part of the set. We built that environment for that particular purpose.

Steina That feedback can also be the spice of the image, the flavouring that you don't really have to see, and it just shapes around whatever object you have; it makes an aura or makes the object more shape-like. And it can also become a mirror effect of whatever object that it repeats into the frame; it's the dimension, it's the space builder. And as you said before, every object that moves affects the feedback, and the feedback affects every object that moves, a mutual manipulation of the real image that's being used, or it can be a synthesized image that's used, and the feedback merging together.

Woody What I would stress the most about feedback is that it itself could lead into all aspects of video; discovering and working with it, it demonstrated all phases of video, yet it may not possess aesthetic quality by itself. What it did to us was to give us the clues to the behaviour of an electronic image, because the sets of clues in the behaviour of feedback are so obvious and so explicit that if you have the imagination to extend that clue into the expression, then you have material which you can learn to control. Of course, the control of feedback is a painful process; it may become frustrating because it is somehow always the same, and somehow always has a similar development, but if you don't really depend on it, if you know how to control it you can really go very far away from the basic. You can use just the flavour of it, just the brilliance of it, just take the cream and leave the garbage on the street. Just bring home the pearl.

What's beautiful about feedback again is that it's also the junk which can generate the beauty; it's the abundance, it's the clay again. Clay is so unattractive unless you bring it home and make something of it. The whole myth of feedback, the put-down or the glorification, is totally meaningless. It provides the vehicle. It's like a drug. It just gives you the ability of seeing what you can expand into. [...]

Steina and Woody Vasulka, extract from 'Woody's Famous Feedback Rap' (a dialogue between the Vasulkas and Jud Yalkut, 1973), in Jud Yalkut, *Electronic Zen: The Alternate Video Generation* (unpublished, 1984) 28–9.

Frank Gillette
Notes for a Proposal on Conceptual Gaming//1973

1. 'Trouble arises', writes Gregory Bateson, 'precisely because the "logic" of adaptation is a different 'logic' from that of the survival and evolution of the ecological system'.[1] The purpose (goal, object, context) of the game is one of simulating ecologic and behavioural complexity ... of distinguishing the sets of relationships between, and the channels of influence exchanged by conceptions of the world and their subsequent control over behaviour in the world.

2. The game is played by 3, 6, 9, 12, 15 or 18 people with a computer system which provides the constantly evolving context within which conceptual models are created and embodied in a range of media, from diagrammatic print-out to holographic simulation. The system also provides the criteria by which models are tested.

3. A primary function of the game is the development of a variety of world-process orientations articulated or embodied in more and more encompassing contexts.

4. How does the game evolve models which separate the contingencies of economic and social behaviour from the bionomic contingences of the ecologic system in which the given behaviour is a constituent part?

5. How does the game evolve corresponding values governed by a meritocracy of ecological description?

6. How does the game separate mythical attitudes based upon the successful domination of nature from conceptions based upon the successful interaction with natural forces?

7. LEXICAL POINTS OF DEPARTURE:

Sequential	Simultaneous, Topological
Linear	Atemporal
Historical	Ahistorical
Labour	Play
Acquisition	Access

Product, Goal	Process
Dualistic	Systemic
Continuity	Discontinuity
Environmental Exploitation	Environmental Enhancement
Ideological	Ecological
Static Image	Moving Image
Taxonomic	Symbiotic, Shared Dependence
Maximum	Optimum
Money	Information

8. Michael Apter[2] pictures the structure of cybernetics thus:

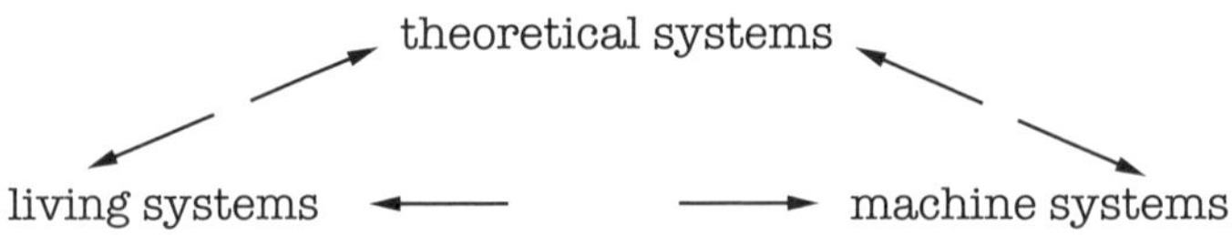

How does the game reflect the interactive flux between these structural elements?

1 [Gregory Bateson, *Steps to an Ecology of Mind* (New York: Ballantine Books, 1972) 339.]

2 Michael Apter, *The Computer Simulation of Behaviour* (New York: Harper & Row, 1970) 43.

Frank Gillette, 'Notes for a Proposal on Conceptual Gaming', *Radical Software*, vol. 2, no. 5 (Winter 1973) 42–3.

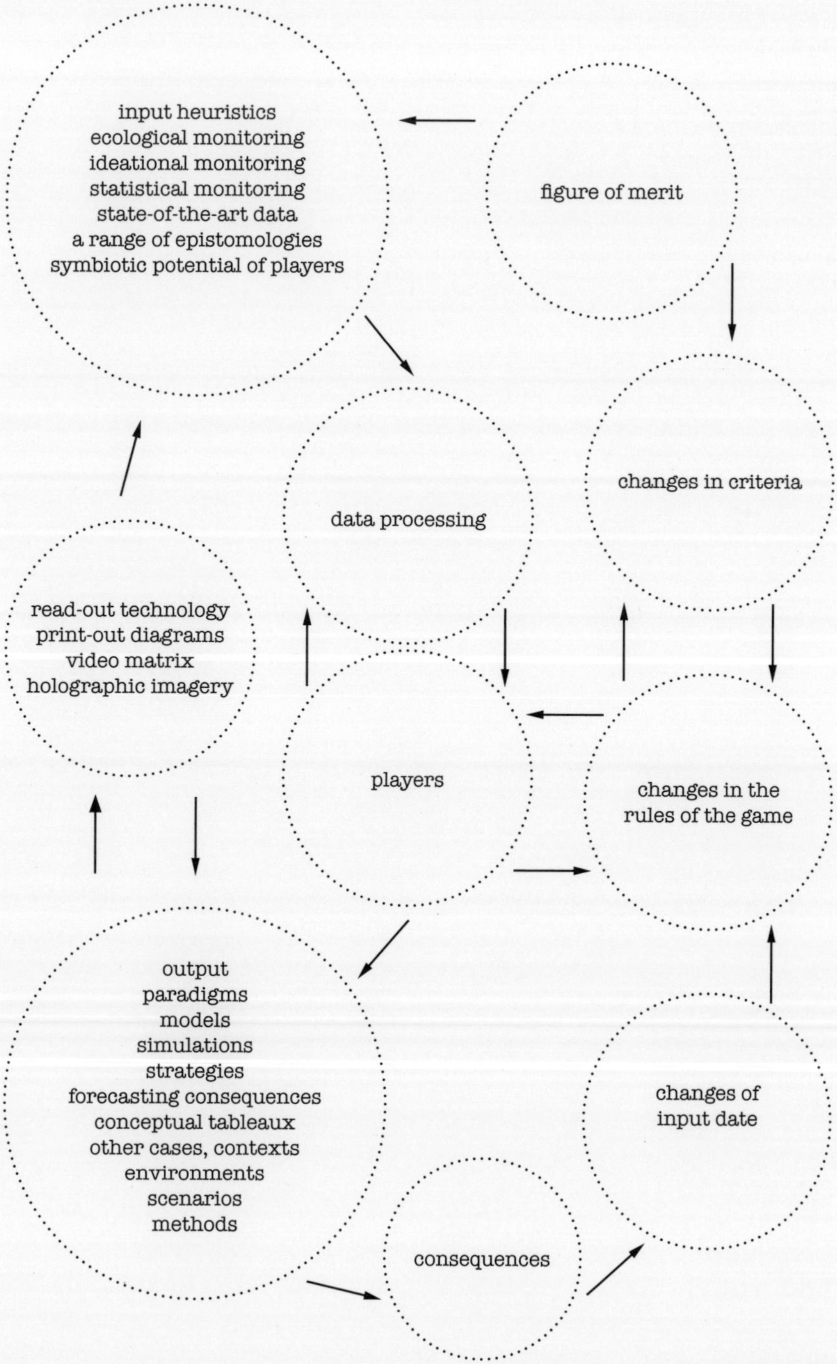
input heuristics
ecological monitoring
ideational monitoring
statistical monitoring
state-of-the-art data
a range of epistomologies
symbiotic potential of players
figure of merit
data processing
changes in criteria
read-out technology
print-out diagrams
video matrix
holographic imagery
players
changes in the
rules of the game
output
paradigms
models
simulations
strategies
forecasting consequences
conceptual tableaux
other cases, contexts
environments
scenarios
methods
changes of
input date
consequences

Zabet Patterson

From the Gun Controller to the Mandala: The Cybernetic Cinema of John and James Whitney//2009

In Los Angeles in the late 1950s, John Whitney started purchasing junk: 'mechanical junk excreted from army depots across the country ... Junk such as brand new thirty-thousand dollar anti-aircraft specialized analogue ballistic problem-solver computers dating back to World War II'.[1] He transformed this military-spec surplus into a machine for creating experimental animation – literally and metaphorically retooling a device that had itself served to remake human vision for modern war. A twin of this machine would enable John's brother James to create the 1966 film *Lapis*, a work P. Adams Sitney would describe as 'the most elaborate example of a mandala in cinema'.[2] [...]

The work began with hand-painted plates. These were used as input for a system that utilized the junked 'hardware from war surplus: selsyn motors to interlock camera functions with artwork motions; ball integrators to preset rate programming of some motions; and differential assemblies to control the incremental advance of the motions as each frame advanced'.[3] As John Whitney later stated, 'it was astonishing to discover the variety of orderly patterns generated by as random a source as these dot patterns. The original artwork contains no hint of the patterns that were produced'.[4] The formal properties of this mandala emerge in dialogue with the apparatus used to create it – the gun controller, a mechanical analogue computer that was a precursor to contemporary computational technology. In this dialogue is crystallized an entire subterranean history of vision and computation that reverberates into the present.

William Moritz situated *Lapis* in an artisanal tradition of experimental film animation that moved from Hans Richter and Viking Eggeling in the 1920s through Len Lye and Oskar Fischinger to the West Coast school with which the Whitneys were involved.[5] Gene Youngblood – writing at much the same time as Moritz – chose to situate *Lapis* in a different trajectory, placing it under the auspices of the human-machine feedback loop. Youngblood's rhetoric is precise; he does not call these works computer films or graphic animations but situates them as 'cybernetic cinema'.[6] In using this terminology Youngblood himself was riding a wave of pop media criticism that began when Buckminster Fuller and Marshall McLuhan a few years prior latched onto cybernetic as a descriptive catchphrase for a diverse range of phenomena from visual media to child psychology. The term's origin, however, can be more precisely located as a theory of messages and information control developed through military research in anti-aircraft technology during and after World War II. During the 1960s, cybernetics had eclipsed these early beginnings to become a more general means of considering and analogizing the organization of bodies – both human and machinic. As such, cybernetics became part of the foundation for an emerging discourse of both human-machine interaction and computational representation. Youngblood's rhetoric situated the Whitneys' films not simply as works made with a computer but as works engaged with this larger field. Youngblood also understood the Whitneys as engaged with the questions the computational turn raised for concepts of representation and visuality and the disciplining of perception in the postwar era. [...]

In the late 1950s, John Whitney continued his experiments in machine-realized art, constructing a machine that he had first imagined during his wartime work at Lockheed. A duplicate of this machine sat in James Whitney's studio.[7] [...] Here, James practised yoga, filmmaking, computer animation, Japanese brush painting and raku pottery, and also studied Taoism and nuclear physics. Here, in the centre, or perhaps off to the side, was the bulky metal apparatus geared with a plethora of moving parts. James Whitney had constructed this machine with his brother John, and he would use it to make a single film – *Lapis*. This large, complex machine with all its gears and selsyn motors, situated at the heart of his studio space, is the pivot around which this story turns; it was a mechanical analogue computer.

[...] Analogue computation implies a representation in which an abstract physical quantity – electric current, light flux – is signified by a concrete physical quantity; for example, length or shape. In this representational practice, an inherent resemblance to the world is maintained. By contrast, digital computing is based upon a rigorous quantization in a practice of unit operations. At the beginning of World War II, a substantial amount of money was spent creating mechanical analogue computers – machines that made their calculations using selsyn gears and wheels and cams. Then, in 1943, the military abruptly replaced these machines with electronic analogue computers. Extraordinarily expensive equipment was dismissed to junk yards by the ton – and John Whitney bought one. The machine he purchased was an M5 anti-aircraft gun director – a special-purpose mechanical analogue computer developed for the guidance and control of anti-aircraft weaponry.[8] These mechanical computers were intricate and elegant integrated systems, each weighing in at approximately 850 pounds and comprising approximately 11,000 moving parts.[9] They were created to solve a particular set of equations for an unchanging number of variables. Specifically, they performed the delicate task of calculating the lead necessary to fire and hit a moving target from a particular distance.[10] [...]

The gun controller is a technology of vision that directly responds to a similar technology of vision. Both are specifically dedicated to augmenting, informing and enframing the soldier's process of seeing – and both directly shape the actions of which he is capable. This is a process in which the human body is re-educated by the machine to act according to a new paradigm of visuality.

At this early stage in the development of computers, the computer was already being developed as a technology that promotes specific habits of visuality. The gun controller trains its users to look at the world in highly specific ways, beginning with glancing – shifting focus back and forth across a visual field. As Jonathan Crary points out, 'one learns nothing new that way; it yields a world already known through habit and familiarity'.[11] One sees an object quickly; then focus, lock, fire. To look at a particular object is to target it. The machine translates the object into data – height, speed, direction – for the singular purpose of burning that object out of the world. To see is to model is to comprehend is to destroy. This would become, in subsequent years, the model for a new kind of visual experience. [...]

This model of vision can be understood more easily in dialogue with the camera obscura than the film camera—and the similarities and differences are instructive. First is a similarity of separation. The individual is held at a certain distance from the world – [...] the operator is enclosed in a sheath of metal and Plexiglas. Like that of the camera obscura, the viewpoint of the turret is a dream of objectivity and transparency, remove and control. But the gun controller operator does not look at a representation of the world held on a separate wall to trace or examine. The eye looks directly on the world, through a framing device. The space of the eye – the space of looking – is shown to be particularly disciplined. Bodily reach is augmented by ballistics. Vision is enframed with lenses. Unlike the camera obscura, where the subject could be said to control or master a world by himself, the subject is himself constrained and enclosed – locked into a circuit of machines. This mode of looking dreams of efficiency and instantaneity; it is a mode appropriate to a space of visual bombardment and a world that valorizes speed.[12] [...]

While contemporary digital computers may seem impossibly distant from these early analogue computers – distinct in size and shape, interface and use – the early computers offer an important reference point for understanding how contemporary computer visuality functions; for example, the tracking of the mouse, and the habits of pointing and clicking. As Crary points out, 'in most cases, using a computer produces a psychic field of expectant attentiveness, within which one inevitably trains oneself to maximize the speed of response to specific commands and functions and in fact to derive at least some satisfaction from these habitual operations of mechanical facility'.[13]

The work of the Whitneys offers us an alternative way of seeing with the computer – but one that responds to, and participates in, the paradigm of vision enacted by the gun controller. As stated earlier, the Whitney brothers both literally and metaphorically retooled the military gun controllers, thereby engaging with a technology that was simultaneously material and social. This technology irrevocably shaped their production – pointing us toward asking how particular computational technologies instantiate particular habits of vision, shaping ideas of subject and object. John Whitney tells us that James Whitney began with random dots. Computer processing repeated, rearranged and recombined these figures, generating precise, strobing patterns. Beginning to watch *Lapis*, the viewer may have an experience like that of the operator of the M5 – an experience that itself worked to develop the habits of vision that Crary attributes to users of modern computers. The eyes seem to defocus. The viewer deliberately relaxes the gaze to take in the whole field at once. The experience is of awaiting motion, then attempting to focus in on its particularity – an attempted targeting, as it were. *Lapis* begins with a glowing sheet of white, a space to project or fall into. Slowly, arrays of tiny particles edge themselves into a ring. The particles swarm and cluster, eventually flaring into complex geometric patterns. Unlike with the M5, this targeting cannot be completed. The image cannot be resolved into a stable emplacement. The viewer is presented with a representation in which there is nothing to lock on to. [...]

Abstract art based on permutation and seriality, as developed in the 1960s, has often been understood as a triumphant celebration of Enlightenment-style reason. As Rosalind Krauss points out, serial geometric abstractions were understood, at this point in time, as 'the demonstration of rationalism itself', the apex of a 'triumphant Cartesianism' that reinscribed the transcendental subject.[14]

This particular debate on abstraction was taking place only shortly after Whitney's initial films were produced – and seems particularly instructive in light of the 'triumph of reason' that the digital computer might seem to represent. Fitting the Whitneys into this debate is fairly straightforward, on one level: the Whitney films are a clear example of algorithmically generated, deterministic abstract art. Further, James Whitney states clearly on a number of occasions that in *Lapis* he wanted to depict 'mind forms'.[15] Yet in spite of the seemingly inherent rationality of these films – they were, after all, systematically worked out by computer – their formal permutations do not emerge as particularly rational. Experientially, these patterns are not easily graspable; nor do they deify the human mind as the site of mathematical prowess. To move inside the systems of the Whitneys' work, including *Permutations* and *Lapis*, is 'precisely to enter a world without a centre, a world of substitutions and transpositions nowhere legitimated by the revelations of a transcendental subject'.[16] [...]

In the years after *Permutations*, John Whitney maintained a belief that the capabilities of computational media would transform art as well as the wider field of human experience. Throughout his lifetime, he continued to dream of the utopian possibilities that computation carried for disrupting traditional modes of representation. James Whitney was less optimistic. Curiously, both brothers' feelings stemmed from a particular aspect of computer technology: its peculiarly dictatorial quality. The M5 was capable of establishing a unique bodily discipline. In a related fashion, *Lapis* produces certain undeniable physical effects (as does *Permutations*). [...]

These effects are not in the film per se but rather in the spectator's perceptual system – 'the electrical-chemical functioning of [the viewer's] own nervous system'.[17] The 'flickering' of film frames can produce strong physiological and psychological effects – including, but not limited to, migraine headaches, nausea, epileptic seizures, anxiety, exhilaration and euphoria. The forceful effect of these films derives from their deft mixing of the purely visual, or optical, with the corporeal, a field that has been described by its phenomenological dimension as 'the haptic'. This mixing took on a particular resonance within the postwar culture of the televisual – as well as within an emerging culture of the computational. [...]

Unlike the distanced reflection of the camera obscura, these new technologies provide no room for distance or judgement – or escape. John Whitney found in this the possibility of a new vision and an accompanying transformation of the human subject. James found disruption, which comes through clearly in the ending to his film *Lapis*. [...]

For James, the computer provoked the sort of physical nausea that can stem only from the furious rejection of a bodily disturbance. Crary has written compellingly of the perceptual retraining that occurred with the advent of capitalism as a dissociation of vision from the body. By contrast, the Whitneys reveal a vision wrought haptic – a reattachment of the body that was understood to hold both revolutionary possibility as well as the danger of a totalizing cybernetic control.

1 John Whitney, *Digital Harmony: On the Complementarity of Music and Visual Art* (Kingsport, Tennessee: Kingsport Press, 1980) 184.

2 P. Adams Sitney, *Visionary Film: The American Avant-Garde, 1943–1978* (Oxford: Oxford University Press, 1974) 264.

3 [footnote 4 in source] Whitney, *Digital Harmony*, op. cit., 184.

4 [5] Ibid., 186.

5 [6] William Moritz, 'Non-Objective Film: The Second Generation', in *Film as Film: Formal Experiment in Film, 1910–1975*, ed. Philip Drummond (London: Hayward Gallery/ Arts Council of Great Britain, 1979) 59–71.

6 [7] Gene Youngblood, *Expanded Cinema* (New York: E.P. Dutton & Co., 1970) 194.

7 [17] John never made a complete work with it, just a catalogue of effects that he would use as a

demo reel. [...] John Whitney's catalogue of effects is now in the collection of the Museum of Modern Art, New York.

8 [19] The M5 anti-aircraft gun director [...] went into mass production in 1940.

9 [20] David Mindell, 'Anti-Aircraft Fire Control and the Development of Integrated Systems at Sperry, 1925–1940', *IEEE Control Systems Magazine*, vol. 15, no. 2 (April 1995) 108–13.

10 [21] The gun director that became known as the M2 director, a predecessor to the M5 and M7, was a complicated mechanical analogue computer that integrated four anti-aircraft guns and an altitude finder into a single, coherent system – albeit one that was dependent on human input.

11 [25] Jonathan Crary, *Suspensions of Perception: Attention, Spectacle and Modern Culture* (Cambridge, Massachusetts: The MIT Press, 1999) 298.

12 [31] See Paul Virilio, *The Vision Machine* (Bloomington: Indiana University Press, 1994) and *War and Cinema: The Logistics of Perception* (London and New York: Verso, 1989).

13 [36] Crary, *Suspensions of Perception*, op. cit., 309.

14 [40] Rosalind Krauss, *The Originality of the Avant-Garde and Other Modernist Myths* (Cambridge, Massachusetts: The MIT Press, 1986) 246.

15 [41] Youngblood, *Expanded Cinema*, op. cit., 222.

16 [42] Krauss, *The Originality of the Avant-Garde*, op. cit., 258.

17 [52] Paul Sharits, 'Notes on Films, 1966–1968', *Film Culture*, no. 47 (1969) 14.

Zabet Patterson, extracts from 'From the Gun Controller to the Mandala: The Cybernetic Cinema of John and James Whitney', *Grey Room*, no. 36 (Summer 2009) 37–54.

Roy Ascott
Art and Telematics: Towards a Network Consciousness//1983

In Mill Valley, California, in the Spring of 1978 I got high on networking. I had anticipated the condition some 17 years earlier in my rather wintry studio in London where I was visited with a cybernetic vision of art, after reading the works of Norbert Wiener and Ross Ashby, formulating a prospectus for creative work which could, as I saw it, raise consciousness to a higher level.

My work, on gallery walls and in colleges of art both in England and abroad (especially at Ealing, London and at the Ontario College of Art [OCA], Toronto) attempted to create analogues of the cybernetic vision which I had committed to publication, but one crucial element was missing. It was not simply that computer access was difficult to arrange, although that certainly was the case at that time, but that some link between the computer and the means of communication was (in my experience) lacking. [...]

More broadly, in my mind, the concept of a global creative network, a cybernetic art matrix, was clear but not until some fifteen years after I had first digested the significance of integrative systems did I come upon the technology which could effect these transformations of culture I had so eagerly anticipated. [...]

In 1980, thanks to an award from the National Endowment for the Arts in Washington DC and Jacques Vallée's Infomedia Notepad computer conferencing system, I set up my first international networking project, mailing portable terminals to a group of artists in California, New York and Wales to participate in collectively generating ideas from their own studios. One of the group, Don Burgy, chose to take his terminal wherever he was visiting and log in from there. [...]

As my contribution to Bob Adrian's 'World in 24 hours' event in Ars Electronica, I had players at their terminals around the world toss coins for the first planetary throw of the I Ching. As I recall we got close to the eighth hexagram, PI (Holding Together/Union), but the bottom line of the lower trigram was unbroken, which transformed the reading into the third hexagram, CHUN (Difficulty at the Beginning): 'Times of growth are beset with difficulties. They resemble a first birth. But these difficulties arise from the very profusion of all that is struggling to attain form. Everything is in motion: therefore if one perseveres there is the prospect of great success.'

Over the past three years I have been interacting through my terminal with artists in Australia, Europe and North America once or twice a week through I.P. Sharp's ARTBOX. I haven't come down from that high yet and frankly I don't expect to. Logging in to the network, sharing the exchange of ideas, propositions, visions and sheer gossip is exhilarating, in fact it becomes totally compelling and addictive. It was Don Burgy, in the first project, whose 26th entry confided: 'Guess I'm hooked because I just got up and the first thing I did (after brushing my teeth) was to log in.'

A new user coming online even for the first time senses a connection and a close community, almost intimacy, which is quite unlike initial face-to-face meetings. For anyone not involved in networking, it is probably hard to imagine how a computer-based medium could possibly be convivial and friendly, or how indeed working at a data terminal could lead to interconnections between human beings at any real level of meaning at all. [...]

[C]omputer-mediated networks, in my view, offer the possibility of a kind of planetary conviviality and creativity which no other means of communication has been able to achieve. One reason may be that networking puts you, in a sense, out of body, linking your mind into a kind of timeless sea. It is a more precise condition than that oceanic feeling that Jung describes in proposing a collective unconscious, and that is because it deals with more than feeling – with particular ideas and associations. These ideas, being generated from a diversity of scattered locations, set in widely different cultural contexts and channelled of course through uniquely different individuals, may become densely layered in meaning and implication. Networking produces an interweaving of imaginations which gives to the term 'associative thinking' the most amplified interpretation. [...]

My belief in this new order of the text, actually a new order of discourse, and my wish to exercise and celebrate the participatory mode of dispersed authorship which networking affords, led me to devise a project wholly concerned with the interweaving of textual inputs from a global distribution of artists. This became *La Plissure du Texte* in the exhibition ELECTRA 1983 at the Musée d'art moderne, Paris.

The title of the project alludes, of course, to Roland Barthes' book *Le Plaisir du Texte*; but pleating (*plissure*) is not intended to replace pleasure (*plaisir*), only to amplify and enhance it. [...] The text Barthes writes about is not 'telematic text' as I experience it, and the authorship he analyses is not the 'distributed authorship' of networking. So that when he celebrates the 'jouissance' that text stimulates, it seems to be very much a solitary act that he describes. Telematic text by contrast, rather than affording a 'jouissance solitaire', offers the means of a 'coming together'. It is a distributed but not dissipated 'jouissance'; metaphysically strange (at first) since the act is indifferent to the geographical location of its contributors, as it is to the time or sequence of their interventions. It constitutes a 'bliss' which is visited on every point of the system which generated it. The processes of coming and going of information are wave-like, and without wishing to stretch the metaphor beyond credibility, at the full intensity of interaction in a creative networking project, these waves can extend to the most prolonged stage of 'jouissance'. [...]

Though we can expect both regional and international regulatory bodies to proliferate in consort with network expansion, the particular nature of telematic discourse makes it less amenable to control. For the artist, its out-of-body, asynchronous, dispersed, interactive and semantically layered qualities make the medium less vulnerable to cultural constraint than earlier modes of expression. [...]

I want to propose, perhaps naively and without caution in the light of society's relentless determination always to institutionalize and contain creativity by any means, that telematic discourse can exist outside such closed systems, or that a much more inclusive, indeed planetary 'fellowship of discourse' can be created, lying outside and circumnavigating the institutional management of discourse as it now exists in book production, conventional telecommunications and entertainment media structures. [...]

Telematics does not only generate a new order of art discourse but demands a new form of criticism and analysis. The theory of this mode of art will have its technical, philosophical and communications aspect bound up within a larger cybernetic framework, which Gregory Bateson has called 'ecology of mind'. This in turn will produce a reevaluation and fresh interpretation of older art forms since it can be argued that meaning has never in reality been created by a one-way dispatch, nor do new ideas of images originate in the solitary mind. Individual genius was the invention of an era, which chose to delimit and contain the subversive power of art within fixed, identifiable boundaries. The field of communications network analysis is especially relevant here, and the major shift of emphasis within this field of research, in recent years, points up the dialectic between the telematic model and the older paradigm of art discourse. [...]

Telematics has arisen from an ethos of cross-disciplinary science and is set within a cybernetic perspective of the world. Numerous writers have attempted to describe the enormous changes they see occurring in human awareness, which some see as a kind of planetary consciousness. Teilhard de Chardin imagined a noosphere, a thinking layer, enveloping the biosphere of the earth. In *The Global Brain* (1983), Peter Russell advanced the hypothesis of the emergence of a planetary brain which may put us onto 'the threshold of a completely new level of evolution, as different from consciousness as consciousness is from life and life is from matter'. He further suggests that this process will result in 'a global brain, which will result in a shift in human ego-centred awareness to a unified field of shared awareness'. [...]

Networking is a shared activity of mind and a form of behaviour which is both a dance and an embrace. It brings about a convergence of ideas from scattered sources which then, amplified, plaited or stacked, diverge out into branching pathways of meaning. This darting to and fro of ideas and images (let's call it creative data), colliding, emitting new combinations, absorbing each other, virtual, real, in a state of continual transformation, puts me in mind of Gary Zukav's description of the dance of sub-atomic particles 'which never ends and is never the same' (*Dancing Wu Li Masters: An Overview of the New Physics*, 1979).

That, I would see as the grand aspiration of networking in art, where the artwork, the transformations of 'creative data,' are in perpetual motion, an unending process. In this sense art itself becomes not a discrete set of entities, but rather a web of relationships between ideas and images in constant flux, to which no single authorship is attributable and whose meanings depend on the active participation of whoever enters the network. In a sense there is one wholeness, the flow of the network in which every idea is a part of every other idea, in which every participant reflects every other participant in the whole. This grand reciprocity, this symmetry of sender and receiver, is such that a mirror image is exchanged in which sender is receiver and receiver sender. The observer of the 'artwork' is a participator who, in accessing the system, transforms it. The physicists who attempt to explain the quantum view that all particles exist potentially as different combinations of other particles often cite the Buddhist parallel view of the world, expressed in the metaphor of Indra's net: 'In the heaven of Indra, there is said to be a network of pearls, so arranged that if you look at one you see all the others reflected in it. In the same way each object in the world is not merely itself but involves every other object and in fact is everything else.'[...]

Roy Ascott, extracts from 'Art and Telematics: Towards a Network Consciousness' (Bristol, 1983), in *Art + Telecommunication*, ed. Heidi Grundmann (Vancouver: The Western Front/Vienna: Blix, 1984) 25–67; reprinted in Roy Ascott, *Telematic Embrace: Visionary Theories of Art, Technology and Consciousness*, ed. Edward A. Shanken (Berkeley and Los Angeles: University of California Press, 2009) 185–200.

Gordon Pask
The Architectural Relevance of Cybernetics//1969

It is easy to argue that cybernetics is relevant to architecture in the same way that it is relevant to a host of other professions; medicine, engineering or law. PERT [Program Evaluation and Review Technique] programming, for example, is unequivocally a 'cybernetic' technique and it is commonly employed in construction scheduling. Computer-assisted design is a 'cybernetic' method and there are several instances of its application to architecture, (for example, the West Sussex County Council's planning scheme in which the designer uses a graphic display to represent the disposition of structural modules on a grid and in which the computer summarizes the cost effect consequences of a proposed layout). Of these cases the first (PERT programming) is a valuable but quite trivial application of cybernetics; the second is likely to have a far-reaching influence upon architectural design. But neither of them demonstrate more than a superficial bond between cybernetics and architecture. If we leave the matter at this level, then architects dive into a cybernetic bag of tricks and draw out those which seem to be appropriate. That is a perfectly reasonable thing to do, of course. But cybernetics and architecture really enjoy a much more intimate relationship; they share a common philosophy of architecture in the sense that Stafford Beer has shown it to be the philosophy of operational research.

The argument rests upon the idea that architects are first and foremost system designers who have been forced, over the last 100 years or so, to take an increasing interest in the organizational (i.e. non-tangible) system properties of development, communication and control. Design problems were coped with as they cropped up, but for some time it has been evident that an underpinning and unifying theory is required. Cybernetics is a discipline that fits the bill in so far as the abstract concepts of cybernetics can be interpreted in architectural terms (and, where appropriate, identified with real architectural systems), to form a *theory* (architectural cybernetics, the cybernetic theory of architecture). [...]

A structure exists chiefly to perform certain functions, for example, to shelter its occupants or to provide them with services. At this level, a 'functional' building is contrasted with a 'decorative' building; it is an austere structure, stripped of excrescences. But the concept of functionalism can be usefully refined in a humanistic direction. The functions, after all, are performed *for* human beings or human societies. It follows that a building cannot be viewed simply in isolation. It is only meaningful as a human environment. It perpetually interacts with its inhabitants, on the one hand serving them and on the other hand controlling their behaviour. In other words, structures make sense as parts of larger systems that include human components and the architect is primarily concerned with these larger systems; *they* (not just the bricks and mortar part) are what architects design. I shall dub this notion architectural 'mutualism', meaning mutualism between structures and men or societies.

One consequence of functionalism and mutualism is a shift of emphasis towards the form (rather than the material constitution) of structures; materials and methods come into prominence quite late in the design process.

Another consequence is that architects are required to design *dynamic* rather than *static* entities. Clearly, the human part of the system is dynamic. But it is equally true (though less obvious) that the structural part must be imaged as continually regulating its human inhabitants.

Once a rudimentary version of the functional/mutualistic hypothesis has been accepted, the integrity of any single system is questionable. Most human/structural systems rely upon other systems to which they are coupled via the human components. By hypothesis, there are organizational wholes which cannot be meaningfully dissected into parts.

Holism is of several types:

a) A functionally interpreted building can only be usefully considered in the context of a city (notice that the city is also functionally interpreted and, as a result, is a dynamic entity).

b) A (functionally interpreted) structure, either a building or an entire city, can only be meaningfully conceived in the context of its temporal extension, i.e. its growth and development.

c) A (functionally interpreted) structure exists as part of an intention, i.e. as one product of a *plan*.

d) If (assumed dogma) man should be aware of his natural surroundings, then buildings should be wedded to or arise from these surroundings ([Frank Lloyd] Wright's organic thesis).

It is a corollary of a, b and c that the structure of a city is not just the carapace of society. On the contrary, its structure acts as a symbolic control programme on a par with the ritual constraints which are known to regulate the behaviour of various tribes and which render this behaviour homeostatic rather than divergent. Hence, the architect is responsible for building conventions and shaping the development of traditions (this comment simply elevates the idea that a building controls its inhabitants to a higher level of organization).

Systems, notably cities, grow and develop and, in general evolve. Clearly, this concept is contingent upon the functionalist/mutualist hypothesis (without which it is difficult to see in what sense the *system* itself *does* grow) though the dependency is often unstated. An immediate practical consequence of the evolutionary point of view is that architectural designs should have rules for evolution built into them if their growth is to be healthy rather than cancerous. [...]

Many human activities are symbolic in character. Using visual, verbal or tactile symbols, man 'talks with' his surroundings. These consist in other men, information systems such as libraries, computers or works of art and also, of course, the structures around him.

Buildings have always been classified as works of art. The novel sub-theory is that structures may be *designed* (as well as intuited) to foster a productive and pleasurable dialogue. [...] Gaudí's work, especially the Parque Guell [is] at a symbolic level one of the most cybernetic structures in existence. As you explore the piece, statements are made in terms of releasers, your exploration is guided by specially contrived feedback, and variety (surprise value) is introduced at appropriate points to make you explore.

It is interesting that Gaudí's work is often *contrasted* with functionalism. Systemically it *is* functionalism pure and simple, though it is aimed at satisfying *only* the symbolic and informational needs of man. [...]

In common with the pure architecture of the 1800s, cybernetics provides a metalanguage for critical discussion. But the cybernetic theory is more than an extension of 'pure' architecture. As we noted somewhat earlier, pure architecture was descriptive (a taxonomy of buildings and methods) and prescriptive (as in the preparation of plans) but it did little to predict or explain. In contrast, the cybernetic theory has an appreciable predictive power. For example, urban development can be modelled as self organizing system (a formal statement of 'evolutionary ideas in architecture') and in these terms it is possible to predict the extent to which the growth of a city will be chaotic or ordered by differentiation. [...]

The cybernetic theory can also claim some explanatory power in so far as it is possible to mimic certain aspects of architectural design by artificial intelligence computer programs (provided, incidentally, that the program is able to learn *about* and *from* architects and by experimenting in the language of architects, i.e. by exploring plans, material specifications, condensed versions of clients' comments, etc.). Such programs are [...] potential aids to design, acting as intelligent extensions of the tool-like programs mentioned at the outset. Further, they offer a means for integrating the constructional system (the 'machinery of production') with the ongoing design process, since it is quite easy to embody the constraints of current technology in a special part of the simulation. However, I believe these programs are of far greater importance as evidencing out theoretical knowledge of what architecture is about. In so far as the program can be written, the cybernetic theory is explanatory.

It seems likely that rapid advances will be made in at least five areas guided by the cybernetic theory of architecture.

1. Various computer-assisted (or even computer-directed) design procedures will be developed into useful instruments.

2. Concepts in very different disciplines (notably social anthropology, psychology, sociology, ecology and economics) will be unified with the concepts of architecture to yield an adequately broad view of such entities as 'civilization', 'city' or 'educational system'.

3. There will be a proper and systematic formulation of the sense in which architecture acts as a social control (i.e. the germ of an idea, mentioned as 'holism', will be elaborated).

4. The high point of functionalism is the concept of a house as a 'machine for living in'. But the bias is towards a machine that acts as a tool serving the inhabitant. This notion will, I believe, be refined into the concept of an environment *with* which the inhabitant cooperates and *in* which he can externalize his mental processes, i.e. mutualism will be emphasized as compared with mere functionalism. [...]

5. Gaudí (intentionally or not) achieved a dialogue between his environment and its inhabitants. [...] The dialogue can be refined and extended [...] in terms of a reactive environment. If, in addition, the environment is malleable and adaptive the results can be very potent indeed. [...]

In the absence of a human inhabitant, feedback leads to stabilization with respect to certain pre-programmed invariants [...] If there is a human being in the environment, the computer, material and all, engages him in dialogue, and within quite wide limits is able to learn about and adapt to his behaviour pattern.

There is thus one sense in which the reactive environment is a *controller* and another in which it is controlled *by* its inhabitants.

In the context of a reactive and adaptive environment, architectural design takes place in several interdependent stages.

1. Specification of the purpose or goal of the system (with respect to the human inhabitants). It should be emphasized that the goal *may* be and nearly always *will* be underspecified [...] [The] aim is to provide a set of constraints that allow for certain, presumably desirable, modes of evolution.

2. Choice of the basic environmental materials.

3. Selection of the invariants which are to be programmed into the system. [...]

4. Specification of what the environment will learn about and how it will adapt.

5. Choice of a plan for adaptation and development. In case the goal of the system is *underspecified* (as in 1) the plan will chiefly consist in a number of evolutionary principles. [...]

Urban planning usually extends over time periods of years or decades and, as currently conceived, the plan is quite an inflexible specification. However, the argument just presented suggests that it need not be inflexible and that urban development could, perhaps with advantage, be governed by a process like that in the dialogue of a reactive environment (physical contact with the inhabitants giving place to an awareness of their preferences and predilections; the inflexible plan to the environmental computing machine). If so, the same design paradigm applies, since in all of the cases so far considered the primary decisions are systemic in character, i.e. they to the delineation or the modification of a control program. This universality is typical of the cybernetic approach.

One final manoeuvre will indicate the flavour of a cybernetic theory. Let us turn the design paradigm in upon itself; let us apply it to the interaction between the designer and the system he designs, rather than the interaction between the system and the people who inhabit it. The glove fits, almost perfectly in the case when the designer uses a computer as his assistant. In other words, the relation 'controller/controlled entity' is preserved when these omnibus words are replaced either by 'designer/system being designed' or by 'systemic environment/inhabitants' or by 'urban plan/city'. But notice the trick: the designer is controlling the construction of control systems, and consequently design is control *of* control, i.e. the designer does much the same job as his system, *but* he operates at a higher level in the organizational hierarchy.

Further, the design goal is nearly always underspecified and the 'controller' is no longer the authoritarian apparatus which this purely technical name commonly brings to mind. In contrast the controller is an odd mixture of catalyst, crutch, memory and arbiter. These, I believe, are the dispositions a designer should bring to bear upon his work (when he professionally plays the part of a controller) and these are the qualities he should embed in the systems (control systems) which he designs.

Gordon Pask, extracts from 'The Architectural Relevance of Cybernetics', *Architectural Design* (September 1969) 494–6 [footnotes not included].

Mary Louise Lobsinger
The Fun Palace Project (1961–64)//2000

[...] Sometime in 1960 Joan Littlewood [a veteran of the English radical theatre scene] met and became friends with Cedric Price [...] a young architect on the London scene. [...]

Price's ideas for a technologically innovative, 'non-deterministic' architecture of planned obsolescence couched in terms of Littlewood's conceptions for alternative theatrical practice produced the quintessential anti-architectural project, the Fun Palace. Littlewood's aesthetic was characterized by an emphasis on direct communication between audience and performer. [It] stressed physical form over speech [...], employed interactive techniques [...], and adapted environmental forms such as festivals with the aim of engaging the sensory and physical participation of the audience in the action. [...] Littlewood's theatrical expertise and social mission were well met by Price's wit and architectural objective: to produce an architecture that could accommodate change. [...]

In 1963 [...] Gordon Pask [...] formed the Committee for the Fun Palace Cybernetic Theatre, which added a new twist to Littlewood's idea of direct communication. With the expertise of an unusual interdisciplinary committee [including Roy Ascott, who proposed an electronic *Pillar of Information*] now in place, the goals of the project were refocused: [...] the technological mandate moved beyond the realm of mechanical mobility into the more ephemeral mobility offered by new information media and mass communications. The discrete disciplinary interests of the three protagonists – cybernetics, transient architecture, participatory theatre and communications – merged in the objectives of the Fun Palace project: to facilitate the emergence of an ephemeral subjectivity through the theatricality of communication. [...]

Although the Fun Palace was never realized, Price achieved such notoriety [...] as to secure for himself a seminal role within debates about architecture and technology. For cutting-edge technological visionaries such as Archigram, Price was the man to watch [...]. The production of the social and the individual – both physically and virtually in real-time – is the theoretical crux of the Fun Palace. [...] The conflict between the simultaneous time of information and the disciplinary time of work (of schedules, timetables, industrial production) had to be amended for humans, to allow them to adapt to the flux and flow of the future technological world. [...]

To facilitate learning and help people live in a scientific culture, the Fun Palace would be equipped with calculating apparatuses (such as cooperative machines operated by two or three people or individual teaching machines) with the idea that these would assist people to learn cooperative behaviour and develop speed in observation and deduction. There would be closed-circuit TVs and surveillance systems by which participants could 'experience the emotional thrill and power' of watching themselves participate [in] a cybernetic learning machine. [...]

Mary Louise Lobsinger, [retitled] extracts from 'Cybernetic Theory and the Architecture of Performance', in *Anxious Modernisms*, ed. Sarah Goldhagan and Rejean Legault (Cambridge, Massachusetts: The MIT Press, 2000) 119–35.

Usman Haque
The Architectural Relevance of Gordon Pask//2007

Gordon Pask (1928–96), English scientist, designer, researcher, academic, playwright, was one of the early proponents and practitioners of cybernetics, the study of control and communication in goal-driven systems of animals and machines. Originally trained as a mining engineer, he went on to complete his doctorate in psychology. His particular contribution was a formulation of second-order cybernetics as a framework that accounts for observers, conversations and participants in cybernetic systems.

Pask was one of the exhibitors at the 'Cybernetic Serendipity' show staged at the ICA, London, in 1968, curated by Jasia Reichardt, an exhibition that became the inspiration for many future interaction designers. The interaction loops of cybernetic systems, such as Pask's *Colloquy of Mobiles* (1968), where actions lead to impacts on the environment that lead to sensing and further modification of actions, are core to the notion of a Paskian environment. He is also known for his Conversation Theory, a particularly coherent and potentially the most productive theory of interaction encompassing human-to-human, human-to-machine and machine-to-machine configurations in a common framework. [...]

The extent of Pask's research, theories and artefact design/construction was enormous. As such, different groups of people find completely different tracts from his back catalogue relevant to their own work. In the 1960s, he worked with the architect Cedric Price on his Fun Palace project as resident cybernetician, introducing the concept of underspecified goals to architecture systems. In the 1970s, Pask's contribution to the philosophy of MIT's Architecture Machine Group was focused around the notion of architecture as an enabler of collaboration. And in the 1980s and early 1990s, architects such as John Frazer at the Architectural Association were particularly interested in how Pask's adaptive systems might be applied to the architectural design process in order to evolve building forms and behaviours.

Now, at the beginning of the 21st century, Pask's Conversation Theory seems particularly important because it suggests how, in the growing field of ubiquitous computing, humans, devices and their shared environments might coexist in a mutually constructive relationship. If we think of having conversations with our environments in which we each have to learn from each other, then Pask's early experiments with mechanical and electrochemical systems provide a conceptual framework for building interactive artefacts that deal with the natural dynamic complexity that environments must have without becoming prescriptive, restrictive and autocratic.

In this context, his teaching and conversational machines demonstrate authentically interactive systems that develop unique interaction profiles with each human participant. This approach contrasts sharply with the 'Star Trek Holodek' approach often attempted in so-called intelligent environments, which presumes that we all see all things in the same way and which denies the creative-productive role of the participant in interactions with such environments. Pask recognized, for example, that interpretation and context are necessary elements in language – as opposed to locating meaning itself in language – which is particularly important to consider for any design process, not least the construction of architectural experience.

His theories on underspecified and observer-constructed goals have been a major influence on my own work. [...] The ongoing projects *Paskian Environments* (with Paul Pangaro, another former student and collaborator of Pask) and *Evolving Sonic Environment* (with Robert Davis, Goldsmiths, University of London) aim to provide concrete and pragmatic strategies for implementing Pask's theories in an architectural context. [...]

Four of Pask's projects in particular give hints about how to create richer, more engaging and stimulating interactive environments. It is worth bearing in mind that each of these predates the common digital computer and was therefore constructed mainly using analogue components. The descriptions below have been simplified, which is somewhat counter to the spirit of a Paskian approach – often necessarily complex – but it is hoped they will provoke the reader to follow up with Pask's own writings, which cover both the theories and the results of the projects he actually constructed.

The *MusiColour Machine*, constructed in 1953, was a performance system of coloured lights that illuminated in concert with audio input from a human performer (who might be using a traditional musical instrument). [...]

The sequence of light outputs might depend at any one moment on the frequencies and rhythms that it can hear, but if the input becomes too continuous – for instance, the rhythm is too static or the frequency range too consistent – *MusiColour* will become bored and start to listen for other frequency ranges or rhythms, lighting only when it encounters those. This is not a direct translation: it listens for certain frequencies, responds and then gets bored and listens elsewhere, produces as well as stimulates improvisation, and reassembles its language much like a jazz musician might in conversation with other band members. Musicians who worked with it in the 1950s treated it very much like another onstage participant.

The innovation in this project is that data (the light-output pattern) is provoked and produced by the participants (other musicians) and nothing exists until one of them enters into a conversation with the designed artefact. In this participant-focused constructional approach, the data evoked has no limits.

Pask constructed a system that aspires to provide enough variety to keep a person interested and engaged without becoming so random that its output appears nonsensical. How these criteria (novelty vs boredom) are measured is core to the system. This calculation is constantly being reformulated on the basis of how the person responds to the response. Unlike the efficiency-oriented pattern-optimization approach taken by many responsive environmental systems, an architecture built on Pask's system would continually encourage novelty and provoke conversational relationships with human participants.

The Self-Adaptive Keyboard Instructor (SAKI), designed by Pask and Robin McKinnon-Wood in 1956, was essentially a system for teaching people how to increase speed and accuracy in typing alphabetic and numeric symbols using a 12-key keyboard.

Whereas contemporaneous teaching machines followed a learn-by-rote model, in which a student attempts to emulate and is then scored for successes, SAKI mimics the possible relationship between a human teacher and student. A teacher is able to respond directly to a student's apparent needs by focusing at times on particular aspects of the material to be studied if weaknesses are measured in these areas. [...]

The result is that, while presentation of test items starts out at the same rate for each item with timely cue information, gradually, as the student improves, the pace is increased and cues are withdrawn for particular items. If a student has difficulty with any individual item – manifested either by making a mistake or by responding slowly – the pace is decreased for that item alone and cue information is selectively reintroduced. [...]

The student responds to the machine just as the machine is responding to the student, and the nature of their goals at any point in time is dependent on the particular history of response the other has provided. [...]

SAKI provides a pragmatic strategy for constructing algorithms that have multiple dynamic environmental inputs and outputs, yet one that is still able to account for an explicitly human contribution. It provides a model of interaction where an individual can directly adjust the way that a machine responds to him or her so that they can converge on a mutually agreeable nature of feedback: an architecture that learns from the inhabitant just as the inhabitant learns from the architecture.

Chemical computers are assemblages constructed electrochemically, that are able to compute an electrical output on the basis of electrical input. In 1958 Pask was particularly interested in how these could be used to construct analogue systems that emulated biological neural networks in their lack of specificity: they evolved behaviours over time depending on how they were trained. Such systems can modify their systemic interconnections as they grow in order to improve proficiency at calculation or pattern recognition. In effect, Pask discovered that they can grow their own sensors. [...]

The fascinating innovation Pask made was to reward the system with an influx of free metal ions – which enable growth of the threads – when certain output criteria were met (as measured at the electrode). The arrangement was so delicate that it was affected by all sorts of inputs including, but not limited to, physical vibration. Though several methods were employed, one in particular is interesting for its potential architectural application as an adaptive environment sensing system. A buzzer was sounded. At the moment of sounding, if the frequency of the buzzer appeared at the sensor electrode, then the system was rewarded with its metal ions. Particular arrangements of thread did occasionally detect the buzzer and replicate the electrical frequency at the sensor electrode.

As a result of the reward system – the provision of metal ions – these types of networks were allowed to survive and prosper while those that did not respond to the buzzer were starved of ions and tended to die off. In other words, by measuring the output criteria (the generated waveform) and rewarding the system when these output criteria correlated with specific input criteria (the buzzer sound), the system became better at recognizing the buzzer. The system was therefore able to evolve its own sound sensor, which would not have been possible if all components of the system had been well specified at the start of the experiment because designing and building such chemical structures would have been prohibitively complex. The underspecification of the threads meant that a much better sound sensor could be evolved and constructed. More importantly though, by changing the input criteria, say by using electromagnetic fields rather than vibration, the system could dynamically grow a new type of sensor.

The reasoning behind Pask's interest in underspecified goals is that if a designer specifies all parts of a design and hence all behaviours that the constituent parts can conceivably have at the beginning, then the eventual identity and functioning of that design will be limited by what the designer can predict. It is therefore closed to novelty and can only respond to preconceptions that were explicitly or implicitly built into it. If, on the other hand, a designed construct can choose what it senses, either by having ill-defined sensors or by dynamically determining its own perceptual categories, then it moves a step closer to true autonomy which would be required in an authentically interactive system. In an environmental sense, the human component of interaction then becomes crucial because a person involved in determining input/output criteria is productively engaging in conversations with his or her environment.

In effect, if such an embodiment has underspecified goals, it enables us to collaborate and converge on shared goals. We are able to affect both the embodiment's response and the way the response is computed.

This is a completely different notion of interaction from that used in many of today's so-called interactive systems, which are premised on unproductive and prespecified circular, deterministic reactions. In these systems, the machine contains a finite amount of information and the human simply navigates through an emerging landscape to uncover it all. I do something, the device/object/environment does something back to me; I do something else, the environment does something else back to me. The human is at the mercy of the machine and its inherent, preconfigured logical system. There is little of the conversation that a truly interactive environment should have, especially in the sense that nothing novel can emerge because all possible responses are already programmed. [...]

Pask was more interested in creating evolving and variable interactions whose sum total is conversational in a valid sense. It is not about concealing and then revealing, but rather about creating information, just as Wikipedia enables in the context of the Web. In an architectural context, this approach enables us to converge, agree on and thereby share each others' conceptual models of a space and what adaptations we decide it requires. With this shared conception we are better able to act upon the givens of a space in conjunction with an artefact, and do so in a constructive, engaging and ultimately satisfying manner. Such a system has to operate with underspecified sensors – either a whole collection of them, each individual sensor of which may or may not eventually be determined as useful in calculating its output and therefore rewarded by the system – or better yet, it may evolve its own sensors, through dynamically determined input criteria.

In his *Colloquy of Mobiles* project (1968), a physically constructed embodiment of Conversation Theory, Pask suspended a collection of purpose-built mechanical artefacts able to move and rotate, some directing beams of light ('females') and others using a combination of servos [feedback-controlled correcting mechanisms] and mirrors to reflect light ('males').

Movement was initially random until a light beam from a female was caught by a male and reflected back to the female's light sensor. At this point, movement would cease and the light beams were locked in place as the males started oscillating their mirrors. After a period of time, the mobiles would start moving again, searching for new equilibrium arrangements.

If left alone, the males and females would continue an elaborate and complex choreography of conversations through the medium of light – one which it was not necessary or even possible to pre-programme – finding coherence every now and then as a light beam was shared between partner members of a conversation. The most interesting point came when visitors entered the scene. Some blocked pathways of light while others used handheld torches to synchronize the devices. The males and females were not able to distinguish between light created by a visitor and light reflected from a female – and had no need to. They were still able to find coherence within their own terms of reference.

Colloquy reminds us that environmental sensor/actuator systems (light beams in this case) will respond to their environment solely on their own terms. [...]

This makes sense for something as easy to learn and understand as a thermostat, in which there is a finite range of input conditions and a finite range of output conditions and the system attempts to map from inputs to outputs in a linear-causal way. However, it becomes problematic in complex environmental systems [...]

Such environmental systems must contain methods for ensuring that proposed outcomes of the system are actually acceptable to the human. The significant complexity and dimensions of the system must be able to improve outcomes without confounding a person with too many inappropriate or incomprehensible outcomes. Moreover, he or she must have a way to reject inappropriateness and reward those criteria that are useful. A person must be able to construct a model of action collaboratively with the environment.

This makes it clear that we need to be able to make coherent connections with our environmental systems. Rather than simply doing exactly what we tell them [...] or alternatively the systems telling us exactly what they think we need (which relies on the environment interpreting our desires, leading to the usual human-machine inequality), a Paskian system would provide us with a method for comparing our conception of spatial conditions with the designed machine's conception of the space.

It is vital at this stage in the development of interactive and time-based media to reconsider Pask's model of interaction, particularly because we are no longer naive in dealing with our technological interfaces. We now expect more from them and are better able to comprehend the structures behind them. A Paskian approach to architecture does not necessarily require complexity of interaction – it relies on the creativity of the person and the machine negotiating across an interface, technological or otherwise. [...]

Architectural systems constructed with Paskian strategies allow us to challenge the traditional architectural model of production and consumption that places firm distinctions between designer, builder, client, owner and mere occupant. Instead we can consider architectural systems in which the occupant takes a prime role in configuring and evolving the space he or she inhabits, a bottom-up approach that enables a more productive relationship with our environments and each other. Pask's approach, if implemented, would provide a crucial counterpoint to the current pervasive computing approach that is founded on interaction loops that have been fixed by the designer and, if implemented, would have a positive impact on the design of future environments.

This interpretation of Pask's way of thinking about interactive systems does not necessarily result in technological solutions. It is not about designing aesthetic representations of environmental data, or improving online efficiency or making urban structures more spectacular. Nor is it about making another piece of high-tech lobby art that responds to flows of people moving through the space, which is just as representational, metaphor-encumbered and unchallenging as a polite watercolour landscape.

It is about designing tools that people themselves may use to construct – in the widest sense of the word – their environments and as a result build their own sense of agency. It is about developing ways in which people themselves can become more engaged with, and ultimately responsible for, the spaces they inhabit. It is about investing the production of architecture with the poetries of its inhabitants.

Usman Haque, extracts from 'The Architectural Relevance of Gordon Pask', *Architectural Design Special Issue: 4Dsocial: Interactive Design Environments*, vol. 77, no. 4 (July/August 2007) 54–61 [footnotes not included].

William J. Mitchell
Intelligent Cities//2007

It's impossible to predict the futures of cities, and certainly unwise to try. For one thing, there are too many uncertainties and random contingencies. For another, there's an indeterminacy effect; interventions concerning the futures of cities – predictions, prophecies, warnings, jeremiads, utopian proposals, science fictions in the style of *Minority Report* [Spielberg, 2002] and the like – themselves have the potential to change thinking and therefore the very futures they address. But designers and planners can usefully suggest *possible* futures, and demonstrate ways to achieve them. This engages the imagination, provides a concrete basis for debate about what might be desirable and achievable, and establishes some starting points for constructive action. In this lecture, then, I will sketch one possible, particularly interesting urban future – that of *intelligent cities*.

Evolution of urban intelligence

To put the idea of intelligent cities in perspective, it is useful to go back to the beginning of a long evolutionary process. The physical fabric of the earliest cities, long before the industrial revolution, consisted essentially of skeleton and skin – columns, beams, walls, floors, and roofs. Its functions were to provide shelter and protection, and to intensify land use. The inhabitants, sometimes assisted by animals, provided their own mobility, performed social and economic transactions face-to-face, and supplied the coordinating intelligence needed to make the city function as a system. This began to establish a cyborg condition; spatially extended layers of artificial skin augmented the protection offered by living human skin. Then, with industrialization, cities started to acquire, as well, increasingly extensive artificial physiologies. Now there were water supply and liquid waste removal networks, energy supply networks, transportation networks, and heating and air-conditioning networks in buildings. Food processing and supply networks extended human alimentary canals at one end, while sewers extended them at the other. Inhabiting a city meant being continually plugged into these networks, and dependent upon them for your survival. Cities extended the capabilities of human bodies in more comprehensive and sophisticated ways, and took over more of the functions traditionally performed by the unaided human body, so the cyborg condition intensified.

Finally, in the latter half of the nineteenth century, cities began to add artificial nervous systems to their fabrics of skeleton, skin, and supply, processing and removal networks. This process began with the construction of telegraph, telephone and radio communication systems, picked up momentum through the first half of the twentieth century, and then accelerated in extraordinary fashion after the introduction of digital telecommunications in the late 1960s – eventually producing today's pervasive connectivity through the internet and mobile wireless networks. The pioneering media theorist Marshall McLuhan presciently hailed these new networks as extensions of human nervous systems.

At the dawn of the twenty-first century, then, cities possessed all of the crucial subsystems of living organisms: structural skeletons; input, processing and waste removal networks for air, water, energy and other essentials; and multiple layers of protective skin. Even more importantly, the existence of artificial nervous systems was enabling cities to sense changes in their internal and external environments and respond, like organisms, in intelligently coordinated fashion. In my 2003 book *Me++: The Cyborg Self and the Networked City*, I discussed this development in detail.

Elements of digital urban intelligence

The elements of artificial urban intelligence did not appear all at once. Instead, there has been a complex and messy process of technological emergence and integration into larger systems – much as, in biological evolution, existing structures and unexpected mutations are appropriated for new purposes within emerging functional organizations. (This sort of process is sometimes called technological convergence, but this terminology suggests something far less messy and ad hoc than what actually goes on.) First came development of the theory of digital information by Claude Shannon, followed in the 1960s by the invention of packet switching, the Arpanet, Ethernet, the Internet and the World Wide Web. Combined with ongoing rapid expansion of wired and wireless communication channels – including very fast fibre optic cable connections – this put in place the necessary nerve pathways at building, city, national and ultimately global scales.

Next, during the 1970s and 1980s, came the increasingly profound effects of the semiconductor revolution. Computers, which had hitherto been large, delicate, expensive and confined to a few specialized and privileged sites became much smaller, much less expensive and much more robust. By the mid 1980s, this development had made desktop personal computers part of everyday life, and these were soon linked into the growing digital networks. With further miniaturization and improvements in performance of semiconductor devices came laptop computers, mobile phones, Blackberries and iPods. Less visibly, but maybe even more importantly in the long run, tiny, embedded microprocessors became crucial components of devices and systems ranging from automobiles to digital cameras. Digital intelligence was no longer tightly concentrated, but was now ubiquitously present throughout urban environments.

During the dotcom bubble of the late 1990s, it seemed to many that the digital era was all about internet connectivity, personal computers and websites. There was much excited discussion (partially but not entirely grounded in reality) of the alleged death of distance, the dematerialization of just about everything, and the emergence of new business opportunities. Meanwhile, though, a third wave of technological innovation – that of digital sensors and tags – was making its presence felt. Minuscule, digital cameras and microphones gave the internet eyes and ears everywhere. GPS and other location technologies made devices such as automobiles and mobile phones continuously aware of where they were. RFID tags embedded in products and packaging began to revolutionize logistics and retailing. All this had the effect of weaving a very tight web of connections between the now-global artificial nervous system and the physical world. The artificial nervous system developed the capacity to perceive and quickly respond to conditions and events in the physical world, while digital processes had increasingly immediate and significant consequences in the physical world. Old metaphors of a distinct 'cyberspace' and transcendent 'virtual worlds' – though still favoured in the popular press, and by some cultural theorists – began to seem quaintly outmoded.

Finally, we have now seen the development of large-scale software that ties all these pieces together to function as intelligently coordinated, geographically distributed systems. The most vivid example of this, of course, is the immense and highly sophisticated software apparatus of Google, which now structures daily intellectual life throughout the world. But there are many more. Today's global financial markets would be impossible without an immense and very sophisticated software infrastructure. Businesses, from financial product manufacturers to airlines, depend upon their enterprise software. Retailers like Amazon.com could not operate at all without the software that manages transactions, keeps track of consumer preferences and handles back-office functions. MySpace and YouTube enable and sustain social and cultural connections through the operation of software. [...] We are also seeing the emergence, in the software world, of cognitive hierarchies similar to those exhibited in the operations of human minds. At the lowest level is software, usually operating on local processors, that provides straightforward, reflex-like capabilities. For example, a sensor-equipped microprocessor in a machine might detect overheating and switch it off. This outage might be noted by central plant management software, which then adjusts the flow of a process accordingly. And this higher-level response, in turn, might be noted and responded to by the still more centralized software for global enterprise management. Such large-scale software systems are now crucial and inescapable in daily urban life. Their economic, social and cultural effects are undeniable, and are increasingly the focus of important social science research. Mostly, I'd be prepared to argue, they have enhanced human life. But they do deserve much closer critical scrutiny – and sometimes resistance – than they have customarily received. They have become very significant expressions of ideology, mediators of consciousness and instruments of power. The new intelligence of cities, then, resides in the increasingly effective combination of digital telecommunication networks (the nerves), ubiquitously embedded intelligence (the brains), sensors and tags (the sensory organs), and software (the knowledge and cognitive competence). This does not exist in isolation from other urban systems, or connected to them only through human intermediaries. There is a growing web of direct connections to the mechanical and electrical systems of buildings, household appliances, production machinery, process plants, transportation systems, electrical grids and other energy supply networks, water supply and waste removal networks, systems that provide life safety and security, and management systems for just about every imaginable human activity. Furthermore, the cross-connections among these systems – both horizontal and vertical – are growing. And we are just at the beginning. [...]

William J. Mitchell, extract from 'Intelligent Cities', *UOC Papers: e-journal on the Knowledge Society*, no. 5 (2007). (www.uoc.edu)

Michael Weinstock
Morphogenesis and the Mathematics of Emergence//2004

Emergence is a concept that appears in the literature of many disciplines, and is strongly correlated to evolutionary biology, artificial intelligence, complexity theory, cybernetics and general systems theory. It is a word that is increasingly common in architectural discourse, where too often it is used to conjure complexity but without the attendant concepts and mathematical instruments of science. In the simplest commonly used definition, emergence is said to be the properties of a system that cannot be deduced from its components, something more than its parts. [...] In the sciences, the term refers to the production of forms and behaviour by natural systems that have an irreducible complexity, and also to the mathematical approach necessary to model such processes in computational environments.

The task for architecture is to delineate a working concept of emergence and to outline the mathematics and processes that can make it useful to us as designers. This means we must search for the principles and dynamics of organization and interaction, for the mathematical laws that natural systems obey, and that can be utilized by artificially constructed systems. We should start by asking: What is it that emerges, what does it emerge from, and how is emergence produced? [...]

It is evident that there is a pressing need for a more developed mathematical approach in current architecture. First, the liberation of tectonics from the economic straitjacket of orthogonal geometry demands more precision in the interface between architectural definitions of form and the computer-driven fabrication processes of manufacturing constructors. Second, the engineering design for the complex geometries of contemporary tectonics must begin from a definitive mathematical base. And third, there is a lacuna in the theoretical body of architecture, an absence that is marked by the proliferation of design processes that borrow the appearance of scientific methods yet lack their clarity of purpose, mathematical instruments and theoretical integrity. [...]

Process and Form

Living organisms can be regarded as systems, and these systems acquire their complex forms and patterns of behaviour through the interactions, in space and over time, of their components. The dynamics of the development of biological forms, the accounts of growth and form, of morphogenesis, have become much more central to evolutionary theory than in Darwin's thesis. [...] Theories of morphogenesis, the creation of forms that evolve in space and over time, are now inextricably entwined with the mathematics of information theory, with physics and chemistry, and with organization and geometry. The pattern of alignment with concepts and technologies of economics and industry remains consistent.

The convergent lines of thought between biology and mathematics were initiated early in the twentieth century, particularly in the work of Whitehead and D'Arcy Thompson. D'Arcy Thompson, zoologist and mathematician, regarded the material forms of living things as a diagram of the forces that have acted on them. His observations of the homologies between skulls, pelvises and the body plans of different species suggested a new mode of analysis, a mathematization of biology. Morphological measurements are specific to species and at times to individuals within a species, and so are various, but there are underlying relations that do not vary – the 'homologies'. [...]

Thompson's comparison of related forms within a genus proceeds by recognizing in one form a deformation of another. Forms are related if one can be deformed into another by Cartesian transformation of coordinates. Comparative analysis reveals what is missing in any singular description of a form, no matter how precise, and that is the morphogenetic tendency between forms.

At around the same time the mathematician and philosopher Whitehead argued that process rather than substance was the fundamental constituent of the world, and that nature consists of patterns of activity interacting with each other. Organisms are bundles of relationships that maintain themselves by adjusting their own behaviour in anticipation of changes to the patterns of activity all around them. Anticipation and response make up the dynamic of life.

The union of these two groups of ideas is very interesting – form and behaviour emerge from process. It is process that produces, elaborates and maintains the form or structure of biological organisms [and non-biological things], and that process consists of a complex series of exchanges between the organism and its environment. Furthermore, the organism has a capacity for maintaining its continuity and integrity by changing aspects of its behaviour. Forms are related by morphogenetic tendencies, and there is also the suggestion that some, if not all, of these characteristics are amenable to being modelled mathematically. The ideas are particularly relevant to us, as in recent years both architecture and engineering have been preoccupied with processes for generating designs of forms in physical and computational environments.

Pattern, Behaviour and Self-Organization

Form and behaviour have an intricate relationship. The form of an organism affects its behaviour in the environment, and a particular behaviour will produce different results in different environments, or if performed by different forms in the same environment. Behaviour is non-linear and context specific.

Mathematical descriptions of behaviour are found in the elaboration of Whitehead's 'anticipation and response' by Norbert Wiener, who developed the first systematic description of responsive behaviour in machines and animals. [...]

Cybernetics organizes the mathematics of responsive behaviour into a general theory of how machines, organisms and phenomena maintain themselves over time. It uses digital and numerical processes in which pieces of information interact and the transmission of information is optimized. Feedback is understood as a kind of 'steering' device that regulates behaviour, using information from the environment to measure the actual performance against a desired or optimal performance.

Work in thermodynamics by Prigogine extended this (and the second law of thermodynamics) by setting up a rigorous and well-grounded study of pattern formation and self-organization that is still of use in the experimental study and theoretical analysis of biological and non-biological systems. He argued that all biological organisms and many natural non-living systems are maintained by the flow of energy through the system. The pattern of energy flow is subject to many small variations, which are adjusted by 'feedback' from the environment to maintain equilibrium, but occasionally there is such an amplification that the system must reorganize or collapse. A new order emerges from the chaos of the system at the point of collapse. The reorganization creates a more complex structure, with a higher flow of energy through it, and is in turn more susceptible to fluctuations and subsequent collapse or reorganization. The tendency of 'self-organized' systems to ever-increasing complexity, and of each reorganization to be produced at the moment of the collapse in the equilibrium of systems, extends beyond the energy relations of an organism and its environment. Evolutionary development in general emerges from dynamic systems.

Geometry and Morphogenesis

Theoreticians fiercely contest the precise relationship of morphogenesis to genetic coding, but there is an argument that it is not the form of the organism that is genetically encoded but rather the process of self-generation of the form within an environment.

Geometry has a subtle role in morphogenesis. It is necessary to think of the geometry of a biological or computational form not only as the description of the fully developed form, but also as the set of boundary constraints that act as a local organizing principle for self-organization during morphogenesis. Pattern and feedback are as significant in the models of morphogenesis as they are in the models of cybernetics and dynamic systems. Alan Turing put forward a hypothesis of geometrical phyllotaxis, the development of form in plants, which offered a general theory of the morphogenesis of cylindrical lattices. These are formed locally rather than globally, node by node, and are further modified by growth. To mathematically model this process, it is necessary to have a global informing geometry, the cylinder, and a set of local rules for lattice nodes. [...]

An intricate choreography of geometrical constraints and geometrical processes is fundamental to self-organization in biological morphogenesis. Computational models of morphogenetic processes can be adapted for architectural research, and self-organization of material systems is evidenced in physical form-finding processes.

The Dynamics of Differentiation and Integration

Feedback is not only important for the maintenance of form in an environment; it is also a useful concept in modelling the relationship of geometrical pattern and form during biological morphogenesis. In pattern form models, feedback is organized in two loops: from form to pattern and from pattern to form. In these models the unstructured formation of biochemical pattern causes morphogenetic 'movements' and a consequent transformation in geometry. The change in geometry disrupts the pattern and a new pattern emerges, which initiates new morphogenetic movements. The process continues until the distribution of morphogens is in equilibrium with the geometry of the evolving form in the model. The feedback loops, from pattern to form and from form to pattern, construct a mathematical model evidenced by the frequency of the terms 'sciences of complexity' and 'complex adaptive systems' in the extensive literature of thermodynamics, artificial intelligence, neural networks and dynamical systems. Mathematically, too, there are commonalities in the approach to computational modelling and simulations. It is axiomatic in contemporary cybernetics that systems increase in complexity, and that in natural evolution systems emerge in increasing complexity, from cells to multicellular organisms, from humans to society and culture.

System theory argues that the concepts and principles of organization in natural systems are independent of the domain of any one particular system, and contemporary research tends to concentrate on 'complex adaptive systems' that are self-regulating. What is common to both is the study of organization, its structure and function. Complexity theory formalizes the mathematical structure of the process of systems from which complexity emerges. It focuses on the effects produced by the collective behaviour of many simple units that interact with each other, such as atoms, molecules or cells. The complex is heterogeneous, with many varied parts that have multiple connections between them, and the different parts behave differently, although they are not independent. Complexity increases when the variety (distinction) and dependency (connection) of parts increases. The process of increasing variety is called differentiation, and the process of increasing the number or the strength of connections is called integration. Evolution produces differentiation and integration in many 'scales' that interact with each other, from the formation and structure of an individual organism to species and ecosystems.

The Genetics of Collective Behaviour

The collective behaviour of semi-autonomous individual organisms is exhibited in the social or group dynamics of many natural species. Flocks of birds and schools of fish produce what appears to be an overall coherent form or array, without any leader or central directing intelligence. Insects such as bees and termites produce complex built artefacts and highly organized functional specializations without central planning or instructions. Structured behaviour emerges from the repetition and interaction of simple rules. Mathematical models have been derived from natural phenomena, massively parallel arrays of individual 'agents', or 'cell units' that have very simple processes in each unit, with simple interactions between them. Complex patterns and effects emerge from distributed dynamical models. Wolfram's extensive study of cellular automata offers a comprehensive account of their characteristics and potential. The study and simulation of co-evolution and co-adaptation is particularly effective in distributed models. [...]

The concepts and mathematical techniques to produce collective behaviour from simple local responses have the potential radically to change architectural environmental systems. It is evident that the current methods of producing 'smart' buildings with hybrid mechanical systems that are controlled by a remote central computer are inferior conceptually and prone to failure in operation. The self-organizing capabilities of distributed dynamic systems have produced intelligent behaviour in natural organisms and in computational simulations, and await architectural applications. [...]

Architecture and Emergence

In answer to the question: What is it that emerges, what does it emerge from, and how is emergence produced? We can say the following.

Form and behaviour emerge from the processes of complex systems. Processes produce, elaborate and maintain the form of natural systems, and those processes include dynamic exchanges with the environment. There are generic patterns in the process of self generation of forms, and in forms themselves. Geometry has both a local and a global role in the interrelated dynamics of pattern and form in self-organized morphogenesis.

Forms maintain their continuity and integrity by changing aspects of their behaviour and by their iteration over many generations. Forms exist in varied populations, and where communication between forms is effective, collective structured behaviour and intelligence emerges.

The systems from which form emerges, and the systems within individual complex forms themselves, are maintained by the flow of energy and information through the system. The pattern of flow has constant variations, adjusted to maintain equilibrium by 'feedback' from the environment. Natural evolution is not a single system but distributed, with multiple systems co-evolving in partial autonomy and with some interaction. An emergent whole form can be a component of a system emerging at a higher level – and what is 'system' for one process can be 'environment' for another.

Emergence is of momentous importance to architecture, demanding substantial revisions to the way in which we produce designs. We can use the mathematical models outlined above for generating designs, evolving forms and structures in morphogenetic processes within computational environments. Criteria for selection of the 'fittest' can be developed that correspond to architectural requirements of performance, including structural integrity and 'buildability'. Strategies for design are not truly evolutionary unless they include iterations of physical (phenotypic) modelling, incorporating the self-organizing material effects of form finding and the industrial logic of production available in CNC [computer numerical control] and laser-cutting modelling machines.

The logic of emergence demands that we recognize that buildings have a life span, sometimes of many decades, and that throughout that life they have to maintain complex energy and material systems. At the end of their life span they must be dissembled and the physical materials recycled. The environmental performance of buildings must also be rethought. The current hybrid mechanical systems with remote central processors limit the potential achievement of 'smart' buildings. Intelligent environmental behaviour of individual buildings and other artefacts can be much more effectively produced and maintained by the collective behaviour of distributed systems.

We must extend this thinking beyond the response of any single individual building to its environment. Each building is part of the environment of its neighbours, and it follows that 'urban' environmental intelligence can be achieved by the extension of data communication between the environmental systems of neighbouring buildings. Urban transport infrastructure must be organized to have similar environmental responsive systems, not only to control internal environments of stations and subways, but also to manage the response to the fluctuating discharge of people onto streets and into buildings. Linking the response of infrastructure systems to groups of environmentally intelligent buildings will allow higher-level behaviour to emerge.

We are within the horizon of a systemic change, from the design and production of individual 'signature' buildings to an ecology in which evolutionary designs have sufficient intelligence to adapt and to communicate, and from which intelligent cities will emerge.

Michael Weinstock, extracts from 'Morphogenesis and the Mathematics of Emergence', *Architectural Design*, special issue: *Emergence: Morphogenetic Design Strategies*, ed. Michael Hensel, Achim Menges, Michael Weinstock (May 2004) 11–17 [footnotes not included].

An artist is not an isolated system. In order to survive he has continuously to interact with the world around him. Theoretically, there are no limits to his involvement

Hans Haacke, Statement, April 1968

SYSTEMS AESTHETICS

Jack Burnham
Systems Aesthetics//1968

A polarity is presently developing between the finite, unique work of high art, i.e. painting or sculpture, and conceptions which can loosely be termed 'unobjects', these being either environments or artefacts which resist prevailing critical analysis. This includes works by some primary sculptors [minimalists] (though some may reject the charge of creating environments), some gallery kinetic and luminous art, some outdoor works, happenings and mixed media presentations. Looming below the surface of this dichotomy is a sense of radical evolution which seems to run counter to the waning revolution of abstract and non-objective art. The evolution embraces a series of absolutely logical and incremental changes, wholly devoid of the fevered iconoclasm which accompanied the heroic period from 1907 to 1925. As yet the evolving aesthetic has no critical vocabulary so necessary for its defence, nor for that matter a name or explicit cause.

In a way this situation might be likened to the 'morphological development' of a prime scientific concept – as described by Thomas Kuhn in *The Structure of Scientific Revolutions* (1962). Kuhn sees science at any given period dominated by a single 'major paradigm'; that is, a scientific conception of the natural order so pervasive and intellectually powerful that it dominates all ensuing scientific discovery. Inconsistent facts arising through experimentation are invariably labelled as bogus or trivial – until the emergence of a new and more encompassing general theory. Transition between major paradigms may best express the state of present art. Reasons for it lie in the nature of current technological shifts.

The economist J.K. Galbraith has rightly insisted that until recently the needs of the modern industrial state were never served by complete expression of the aesthetic impulse. Power and expansion were its primary aims.

Special attention should be paid to Galbraith's observation. As an arbiter of impending socio-technical changes his position is pivotal. For the Left he represents America's most articulate apologist for Monopoly Capitalism; for the Right he is the socialist *eminence grise* of the Democratic Party. In *The New Industrial State* (1967) he challenges both Marxist orthodoxies and American mythologies premised upon *laissez-faire* capitalism. For them he substitutes an incipient technocracy shaped by the evolving technostructure. Such a drift away from ideology has been anticipated for at least fifty years. Already in California thinktanks and in the central planning committees of each soviet, futurologists are concentrating on the role of the technocracy, i.e. its decision-making autonomy, how it handles the central storage of information, and the techniques used for smoothly implementing social change. In the automated state, power resides less in the control of the traditional symbols of wealth than in information.

In the emergent 'superscientific culture' long-range decision making and its implementation become more difficult and more necessary. Judgement demands precise socio-technical models. Earlier the industrial state evolved by filling consumer needs on a piecemeal basis. The kind of product design that once produced 'better living' precipitates vast crises in human ecology in the 1960s. A striking parallel exists between the 'new' car of the automobile stylist and the syndrome of formalist invention in art, where 'discoveries' are made through visual manipulation.

Increasingly 'products' – either in art or life – become irrelevant and a different set of needs arise: these revolve around such concerns as maintaining the biological livability of the Earth, producing more accurate models of social interaction, understanding the growing symbiosis in man-machine relationships, establishing priorities for the usage and conservation of natural resources, and defining alternate patterns of education, productivity and leisure. In the past our technologically-conceived artefacts structured living patterns. We are now in transition from an *object-oriented* to a *systems-oriented* culture. Here change emanates, not from *things*, but from *the way things are done*.

The priorities of the present age revolve around the problems of organization. A systems viewpoint is focused on the creation of stable, ongoing relationships between organic and non-organic systems, be these neighbourhoods, industrial complexes, farms, transportation systems, information centres, recreation centres, or any of the other matrixes of human activity. All living situations must be treated in the context of a systems hierarchy of values. Intuitively many artists have already grasped these relatively recent distinctions, and if their 'environments' are on the unsophisticated side, this will change with time and experience.

The major tool for professionally defining these concerns is systems analysis. This is best known through its usage by the Pentagon and has more to do with the expense and complexity of modern warfare than with any innate relation between the two. Systems analysts are not cold-blooded logicians; the best have an ever-expanding grasp of human needs and limitations. One of the pioneers of systems applications, E.S. Quade, has stated that 'Systems analysis, particularly the type required for military decisions, is still largely a form of art. Art can be taught in part, but not by the means of fixed rules ...'[1] Thus 'The Further Dimensions'[2] elaborated upon by Galbraith in his book are aesthetic criteria. Where for some these become the means for tidying up a derelict technology, for Galbraith aesthetic decision-making becomes an integral part of any future technocracy. As yet few governments fully appreciate that the alternative is biological self-destruction.

Situated between aggressive electronic media and two hundred years of industrial vandalism, the long held idea that a tiny output of art objects could somehow 'beautify' or even significantly modify the environment was naive. A parallel illusion existed in that artistic influence prevails by a psychic osmosis given off by such objects. Accordingly lip service to public beauty remains the province of well-guarded museums. Through the early stages of industrialism it remained possible for decorative media, including painting and sculpture, to embody the aesthetic impulse; but as technology progresses this impulse must identify itself with the means of research and production. Obviously nothing could be less true for the present situation. In a society thus estranged only the didactic function of art continues to have meaning. The artist operates as a quasi political *provocateur*, though in no concrete sense is he an ideologist or a moralist. '*L'art pour l'art*' and a century's resistance to the vulgarities of moral uplift have ensured that.

The specific function of modern didactic art has been to show that art does not reside in material entities, but in relations between people and between people and the components of their environment. [...]

In an advanced technological culture the most important artist best succeeds by liquidating his position as artist vis-a-vis society. Artistic nihilism established itself through this condition. At the outset the artist refused to participate in idealism through craft. 'Craft-fetishism',[3] [...] remains the basis of modern formalism. Instead the significant artist strives to reduce the technical and psychical distance between his artistic output and the productive means of society. Duchamp, Warhol and Robert Morris are similarly directed in this respect. Gradually this strategy transforms artistic and technological decision-making into a single activity – at least it presents that alternative in inescapable terms. Scientists and technicians are not converted into 'artists', rather the artist becomes a symptom of the schism between art and technics. Progressively the need to make ultra sensitive judgements as to the uses of technology and scientific information becomes 'art' in the most literal sense.

As yet, the implication that art contains survival value is nearly as suspect as attaching any moral significance to it. Though with the demise of literary content, the theory that art is a form of psychic preparedness has gained articulate supporters.

> Art, as an adaptive mechanism, is reinforcement of the ability to be aware of the disparity between behavioural pattern and the demands consequent upon the interaction with the environment. Art is rehearsal for those real situations in which it is vital for our survival to endure cognitive tension, to refuse the comforts of validation by affective congruence when such validation is inappropriate because too vital interests are at stake.[4] [...]

The systems approach goes beyond a concern with staged environments and happenings; it deals in a revolutionary fashion with the larger problem of boundary concepts. In systems perspective there are no contrived confines such as the theatre proscenium or picture frame. Conceptual focus rather than material limits define the system. Thus any situation, either in or outside the context of art, may be designed and judged as a system. In as much as a system may contain people, ideas, messages, atmospheric conditions, power sources, etc., a system is, to quote the systems biologist, Ludwig von Bertalanffy, a 'complex of components in interaction',[5] comprised of material, energy and information in various degrees of organization. In evaluating systems the artist is a perspectivist considering goals, boundaries, structure, input, output and related activity inside and outside the system. Where the object almost always has a fixed shape and boundaries, the consistency of a system may be altered in time and space, its behaviour determined both by external conditions and its mechanisms of control. [...]

[F]or our time the emerging major paradigm in art is neither an *ism* nor a collection of styles. Rather than a novel way of rearranging surfaces and spaces, it is fundamentally concerned with the implementation of the art impulse in an advanced technological society. As a culture producer, man has traditionally claimed the title *Homo Faber*: *man the maker* (of tools and images). With continued advances in the industrial revolution, he assumes a new and more critical function. As *Homo Arbiter Formae* his prime role becomes that of *man the maker of aesthetic decisions*. These decisions – whether they are made concertedly or not – control the quality of all future life on the Earth. Moreover these are value judgements dictating the direction of technological endeavour. Quite plainly such a vision extends beyond political realities of the present. This cannot remain the case for long.

1 E.S. Quade, 'Methods and Procedures', in *Analysis for Military Decisions* (Santa Monica: Rand Corporation, November 1964) 153.

2 J.K. Galbraith, *The New Industrial State* (Boston: Houghton Mifflin Co., 1967) 343–53.

3 Christopher Caudwell (pseud.), *Illusion and Reality: A Study of the Sources of Poetry* (London: Macmillan, 1937) 111.

4 Morse Peckham, *Man's Rage for Chaos: Biology, Behaviour and the Arts* (New York: Schocken Books, 1965) 314.

5 [footnote 6 in source] Ludwig von Bertalanffy, *Robots, Men and Minds* (New York: George Braziller Inc., 1967) 69.

Jack Burnham, extracts from 'Systems Aesthetics', *Artforum*, vol. 7, no. 1 (September 1968) 30–35.

Jack Burnham
Real Time Systems//1969

Presently it will be accepted that art is an archaic information processing system, characteristically Byzantine rather than inefficient. To emphasize this cybernetic analogy, programming the art system involves some of the same features found in human brains and in large computer systems. Its command structure is typically hierarchical.[1] At the basic level artists are similar to programs and subroutines. They prepare new codes and analyse data in making works of art.

These activities are supervised by metaprograms which consist of instructions, descriptions and the organizational structures of programs. Metaprograms include art movements, significant stylistic trends, and the business, promotional and archival structures of the art world. At the highest level art contains a self-metaprogram which, on a long-term basis, reorganizes the goals of the art impulse. The self-metaprogram operates as an undetected overseer, establishing strategies on all lower levels in terms of societal needs. Because we have no comprehensive picture of human life, these needs remain rather obscure (*Zeitgeist* is not sufficiently teleologic to express the anticipatory monitoring function of the self-metaprogram).

Aesthetic values emanate from the self-metaprogram. [...] Values, though, are simply the result of long term information processing structures. This is the business of museums and art historians. The more aggressive commercial galleries have long considered controlling and creating art information vital to selling, while not forgetting that sales are art information. The survival strategy of all social organizations, including the art system, is that of transforming preferred information into values.

In business this is taken for granted. At the management level, information 'is data that has been culled, analysed, interpreted and presented on a selective basis in a manner useful for understanding and decision making. Its function is to decrease uncertainty'.[2] As indicated, every artist produces data by making art. Critics, magazines, galleries, museums, collectors and historians exist to create information out of unprocessed art data. History is uncertainty about art minimized.

A major illusion of the art system is that art resides in specific objects. Such artefacts are the material basis for the concept of the 'work of art'. But in essence, all institutions which process art data, thus making information, are components of the work of art. Without the support system, the object ceases to have definition; but without the object, the support system can still sustain the notion of art. So we can see why the art experience attaches itself less and less to canonical or given forms but embraces every conceivable experiential mode, including living in everyday environments. Thus art, according to John McHale, becomes 'temporal immersion in a continuous contextual flow of communicated experiences'.[3] Examine the function of information in art: communication theory states that information is obtained when a signal reduces uncertainty within a system. Information is need requited; hence information for a system has high entropy-reducing potential (negentropy). Negentropy is the ability of information to increase the structure and potential energy within a system. Such information is only obtained by expending the energy of systems outside the one receiving information. Thus the art system has maintained its vitality by constantly reaching outside of itself for data. In the past this has taken the form of new subject matter, materials and techniques. But art now challenges the entire art information processing structure, not merely its content.

Encoding information always involves some physical process. In high-speed processing this takes the form of digital computer 'hardware'. The procedures or programs for processing data are called 'software'. For all previous art, distinctions between software and hardware were not recognized, so that encoding took the form of other art media and materials, where some information was lost, and perhaps some gained. Graphic reproductions of original works of art were a form of advertising. We now look upon them as works of art in their own right. Electronics have taught us that we often confuse software with its physical transducer. In other words, if we extend the meaning of software to cover the entire art information processing cycle, then art books, catalogues, interviews, reviews, advertisements, sales and contracts are all software extensions of art, and as such legitimately embody the work of art. The art object is, in effect, an information 'trigger' for mobilizing the information cycle. Making, promoting and buying art are real time activities. That is to say, they happen within the day-to-day flow of normal experience. Only Art Appreciation happens in ideal, non-existential time. [...]

In societies where existing values adequately deal with the environment, there are no comparative values – only the existing way of life. Values are non-existent in metabolically stable societies. Hopefully such a metabolic reorganization is under way and will lead to a convergence of global information structures with parallel rather than linear processing. [...]

Objectively, we know very little about the rules of this metabolism. But we know that organic stability is predicated upon extensive communication networks, including memory, feedback and automatic decision-making capacities. The rudiments of such networks already exist, in the form of large-scale digital computer control systems. SAGE, the first computer-based air defence system; Project Mercury, the first real time digital support system for space flight; Telefile, the first online banking system; and SABRE, the first computerized airline reservation system, are a few of many operating real time systems which gather and process data from environments, in time to effect future events within those environments.

Emotionally most humanists share an instinctive antipathy for these immensely complex computer systems. Their Orwellian overtones far overshadow their conceivable use as artists' tools. But practically, it is imperative that artists do understand them – both technically and philosophically. These computer systems deal with *real time* events, events which are uncontrived and happen under normal circumstances. All of the data processing systems I have referred to are *built into and become a part of* the events they monitor. Already a large part of the metabolic information used to run the military and commercial interests of the United States is real time-oriented. It is not proposed that artists have the choice between traditional media or using the computer. What I am saying is that the real time information-processing mode is rapidly becoming the routine style of handling information. [...]

Some recent tendencies in Hans Haacke's work intrigue me. One is a willingness to use all forms of organic life – from the most elementary to the most complicated. This seems a logical extension of his philosophy of natural systems. [...] For a museum, he is planning a steady output of statistical information about visitors involving a small process-control computer and a display device. Two years ago Haacke would have balked at using this kind of technology; today, working more closely with events, it becomes a necessity. As Haacke explains:

> The artist's business requires his involvement in practically everything ... It would be bypassing the issue to say that the artist's business is how to work with this and that material or manipulate the findings of perceptual psychology, and that the rest should be left to other professions ... The total scope of information he receives day after day is of concern. An artist is not an isolated system. In order to survive ... he has continuously to interact with the world around him. Theoretically, there are no limits to his involvement.[4] [...]

Art as information-processing leaves little in the way of protection for the artist. Style used to be the art system's equivalent to patent rights. And even among the conceptualists one senses a certain degree of deference and respect for each other's ideas. But if the output of artists continues to be based upon non-sequential ideas, it may be impossible to support the notion of 'ownership'. Such ownership amounts to who amplifies original data first so that it becomes information. [...] According to [the ecologist Ramon] Margalef, boundaries between systems in nature are usually asymmetrical. More organized systems always gain information and energy from less organized systems. [...]

> It is a basic property of nature, from the point of view of cybernetics, that any exchange between two systems of different information content does not result in a partition or equalizing of the information, but increases the difference. The system with more accumulated information becomes still richer for the exchange. Broadly speaking, the same principle is valid for persons and human organizations: any exchange increases to a greater extent the information of the party already better informed.[5]

Little imagination is needed to realize how this principle operates within the art system. As the fame of a living artist grows, he ceases simply to make data make data. His subsequent output is information, since it is already art history. Plagiarism of existing information, i.e. the work of well-known artists, has minimal energy – unless original information becomes the object of new data in a very convincing way. On the other hand, famous 'avant garde' artists may capitalize upon the work of their lesser-known contemporaries. Being better organized systems, established artists have greater access to museums and media. It is important, however, that they use such material while it is still data, i.e. before it becomes art information.

On a personal level Margalef's cybernetic principle remains a matter of ethics and practicality. But its implications for the total art information system are far-reaching. As information processing becomes more generally understood, institutions and persons – other than artists – will insist upon creating their own art information. Specifically I am thinking about projects which demand money, planning and technical support far beyond the individual artist's means. Artistic endeavour is thus brought up (or down) to the level of corporate research. [...]

1 John Cunningham Lilly, M.D., *Programming and Metaprogramming in the Human Biocomputer: Theory and Experiments* (Baltimore: Communication Research Institute, 1967).

2 Carl Heyel, *Computers, Office Machines and the New Information Technology* (New York: Macmillan, 1969) 178.

3 John McHale, *The Future of the Future* (New York: George Braziller Inc., 1969) 300.

4 [footnote 9 in source] From a talk by Hans Haacke at the Annual Meeting of the Intersocietal Color Council, April 1968.

5 [16] Ramon Margalef, *Perspectives in Ecological Theory* (Chicago: University of Chicago Press, 1968) 16–17.

Jack Burnham, extracts from 'Real Time Systems', *Artforum*, vol. 8, no. 1 (September 1969) 49–55.

Hans Haacke
In Conversation with Jeanne Siegel//1971

Jeanne Siegel You have been called a naturalist because of your extensive interest in physical elements, as well as grass, birds, ants and animals.

Hans Haacke [...] When people see the wind stuff or the things I have done with animals, they call me a 'naturalist'. Then they get confused or feel cheated when they discover, for example, my interest in using a computer to conduct a demographic survey. This is inconsistent only for those with a naive understanding of nature – nature being the blue sky, the Rockies, Smokey the Bear. The difference between 'nature' and 'technology' is only that the latter is man-made. The functioning of either one can be described by the same conceptual models, and they both obviously follow the same rules of operation. It also seems that the way social organizations behave is not much different. The world does not break up into neat university departments. It is one supersystem with myriad subsystems, each one more or less affected by all the others.

If you take a grand view, you can divide the world into three or four categories – the physical, biological, the social and behavioural – each of them having interrelations with the others at one point or another. There is no hierarchy. All of them are important for the upkeep of the total system. It could be that there are times when one of these categories interests you more than another. So, for example, I now spend more thought on things in the social field, but simultaneously I am preparing a large water-cycle for the Guggenheim show that uses the peculiarities of the building.

Siegel When did you first become aware of systems theory?

Haacke Sometime in 1965 or 66 I was introduced to the concept of systems. I heard about systems analysis, and the related fields of operational research, cybernetics, etc. The concepts used in these fields seemed to apply to what I had been doing, and there was a useful terminology that seemed to describe it much more succinctly than the terminology that I and other people had been using until then, so I adopted it. But using a new terminology doesn't mean that the work described has changed. A new term is nothing holy, so it can't serve as a union label. On the other hand, a clear terminology can help to stimulate thinking.

Siegel Jack Burnham has had a lot to say about systems and sculpture, yours in particular. When did you first meet him?

Haacke I met Jack in 1962 when we were both isolated from people interested in what we were doing. Since then we have been in contact and have had a very fruitful exchange of ideas. It was Jack who introduced me to systems analysis.

Siegel What is your definition of a system that is also a work of art?

Haacke A system is most generally defined as a grouping of elements subject to a common plan and purpose. These elements or components interact so as to arrive at a joint goal. To separate the elements would be to destroy the system. The term was originally used in the natural sciences for understanding the behaviour of physically interdependent processes. It explained phenomena of directional change, recycling and equilibrium. I believe the term system should be reserved for sculptures in which a transfer of energy, material or information occurs, and which do not depend on perceptual interpretation. I use the word 'systems' exclusively for things that are not systems in terms of perception, but are physical, biological, or social entities which, I believe, are more real than perceptual titillation. [...]

A very important difference between the work of Minimal sculptors and my work is that they were interested in inertness, whereas I was concerned with change. From the beginning the concept of change has been the ideological basis of my work. All the way down there's absolutely nothing static – nothing that does not change, or instigate real change. Most Minimal work disregards change. Things claim to be inert, static, immovably beyond time. But the status quo is an illusion, a dangerous illusion politically. [...]

Siegel Is there any difference in communication between social systems and physical or biological ones?

Haacke For physical or biological processes to take their course, there is no need for the presence of a viewer – unless, as with some participatory works, his physical energy is required (he then becomes an indispensable part of the system's physical environment). However, there is no need for *anybody* to get mentally involved. These systems function on their own, since their operation does not take place in the viewer's mind (naturally this does not prevent a mental or emotional response).

The rigging of a social situation, however, usually follows a different pattern. There the process takes place exclusively in the minds of people. Without participants there is no social set. Take the *MoMA Poll* in last year's 'Information' show: the work was based on a particular political situation circumscribed by the Indochina War, Nixon's and Rockefeller's involvement in it. MoMA's close ties to both, my own little quarrels with the museum as part of the Art Workers Coalition's activities, and then all the minds of the people who had a stake in this game – the Vietcong as much as the Scarsdale lady on her culture tour to the city. The result of the poll – approximately 2 to 1 against Rockefeller/Nixon and the war – is only the tip of the iceberg. The figures are not quite reliable because MoMA, as usual, did not follow instructions, and the polls have to be taken with a grain of salt.

Emily Genauer gave us a little glimpse of the large base of the work in her review of the show. She wrote: 'One may wonder at the humour (propriety, obviously, is too archaic a concept even to consider) of such poll-taking in a museum founded by the governor's mother, headed now by his brother, and served by himself and other members of his family in important financial and administrative capacities since its founding 40 years ago.' With this little paragraph she provided some of the background for the work that was not intelligible for the politically less-informed visitors of the museum. She also articulated feelings that are shared by the top people at numerous museums. It goes like this: We are the guardians of culture. We honour artists by inviting them to show in *our* museum, we want them to behave like guests; proper, polite and grateful. After all, we have put up the dough for this place.

The energy of information interests me a lot. Information presented at the right time and in the right place can be potentially very powerful. It can affect the general social fabric.

Such things go beyond established high culture as it has been perpetrated by a taste-directed art industry. Of course I don't believe that artists really wield any significant power. At best, one can focus attention. But every little bit helps. In concert with other people's activities outside the art scene, maybe the social climate of society can be changed. Anyway, when you work with the 'real stuff' you have to think about potential consequences. A lot of things would never enter the decision-making process if one worked with symbolic representations that have to be weighed carefully. If you work with real-time systems, well, you probably go beyond Duchamp's position. Real-time systems are double agents. They might run under the heading 'art', but this culturization does not prevent them from operating as normal. The *MoMA Poll* had even more energy in the museum than it would have had in the street – real socio-political energy, not awe-inspiring symbolism. [...]

Hans Haacke and Jeanne Siegel, extracts from 'An Interview with Hans Haacke', *Arts* magazine, vol. 45, no. 7 (May 1971) 18–21.

Edward A. Shanken
Reprogramming Systems Aesthetics//2009–14

As the cult of high modernism tumbled from its lofty throne, the scientific theories of Claude Shannon, Norbert Wiener and Ludwig von Bertalanffy gained substantial purchase in the arts. Radically opposed to the romantic emotionality of expressionism, Abraham Moles and Max Bense's theories of 'information aesthetics', Roy Ascott's cybernetic art theories and Jack Burnham's 'systems aesthetics' became influential models for more rational approaches to making and understanding art. Losing their lustre by the mid 1970s, they disappeared from art discourses for nearly two decades, apparently gathering dust but, as recent affairs suggest, also gathering steam. Historical and critical writing addressing these aesthetic theories began to emerge in the 1990s and accelerated in the 2000s, when a number of exhibitions and symposia were devoted to related themes (including a 'Systems Art' symposium at the Whitechapel Gallery in 2007). Specialized scholarly publications also mushroomed in the 2000s, including Francis Halsall's *Systems of Art* (2008). Paralleling the entry of this historical recuperation into museum contexts, scholarly writing on the subject has entered into more mainstream academic discourses, including Pamela Lee's *Chronophobia* (MIT Press, 2004) and the celebration of Burnham's work in the fiftieth anniversary issue of *Artforum* in 2012. To borrow a line from Hans Haacke's proposed 1971 work ironically dedicated to Wiener and resuscitated by scholar Luke Skrebowski: 'All Systems Go!'[1]

Contemporary discourses surrounding systems aesthetics, however, tend to lack an appreciation of the alternate art histories that emerged around informational, cybernetic and systems approaches to art. Charlie Gere identifies early conceptions of systems thinking and computation applied to art in the Independent Group's exhibition catalogue for *This is Tomorrow* (Whitechapel Gallery, 1956) and notes John McHale's 1962 pronouncement that 'the future of art seems no longer to lie with the creation of enduring masterworks but with defining alternative cultural strategies, through a series of communicative gestures in multi-media forms'.[2] Roy Ascott wrote about the application of cybernetics to art in 1963, proposed human-machine symbiosis as art in 1964, anticipated remote interdisciplinary collaborations involving artists in 1966–67, and in 1967 proclaimed, 'When art is a form of behaviour, software predominates over hardware in the creative sphere. Process replaces product in importance, just as system supersedes structure', all foundations undergirding his subsequent praxis of telematic art.[3] In 2006 [in *Materializing New Media*] Anna Munster proposed 'information aesthetics' as a 'new kind of aesthetics', apparently unaware of Bense and Moles' theorizations of the late 1950s using the same term, and seemingly equally oblivious to Burnham's systems aesthetics. So, while it is important to recognize the vital contributions of Burnham's theories, it is equally important to recognize that they were not without precedent, and that those precedents contributed to the overall ecology of discourses of which his were a part, just as recent scholarship on systems aesthetics is part of a larger ecology of art-historical writing. The emerging literature has only begun to scrutinize these issues and to contend with why those aesthetic theories lost artistic currency in the 1970s, how they increasingly and differentially came to regain it, beginning in the 1990s, and what their possible hermeneutic uses are today. The question I propose is: How has the historicization of those interpretive syntheses in the 1960s been 'reprogrammed' by contemporary artists and writers, and to what ends?

Marga Bijvoet's *Art as Inquiry: Toward New Collaborations Between Art, Science and Technology* (1997) is a pioneering yet under-recognized monographic study of art in the 1960s and early 1970s.[4] A key aspect of Bijvoet's framing of this terrain draws on information theory, cybernetics and systems theory, with particular emphasis on the aesthetic theories of Jack Burnham. She discusses the application of biologist Ludwig von Bertalanffy's general system theory in Burnham's formulation of a 'systems aesthetics' in his *Artforum* essay of that title and in his book *Beyond Modern Sculpture*, both published in 1968.[5]

In 'Gemini Rising, Moon in Apollo' (1998), I noted that in presenting 'such diverse artists as Joseph Kosuth, Hans Haacke and Sonia Sheridan', Burnham's 1970 'Software' exhibition, 'implicitly problematized distinctions between "art and technology" and other experimental art media and technological invention' including what had come to be known as hypertext and intelligent environments.[6] In 'The House that Jack Built' (1998), I claimed that the relationship Burnham posited 'between experimental art practices and "art and technology" questioned conventional distinctions between them, and offered important insights into the complementarity of conventional, experimental and electronic media in the emerging cultural paradigm later theorized as postmodernity'.[7]

Mitchell Whitelaw's 1998 essay, '1968/1998: Rethinking a Systems Aesthetic'[8] emphasized Burnham's 'anticipation of contemporary concerns', such as the 'cybernetic organism', 'self-organizing systems in relation to sculpture', and 'an art embracing "real time information processing"'. Similarly, he noted, the re-entry of terms like cybernetics and systems into the critical vocabulary of cultural discourse give new relevance to Burnham's systems aesthetics.

Simon Penny states that he gravitated to Burnham's 'visionary and pioneering' writing as a sculpture student in the late 1970s, and notes that it influenced his pursuit of interactive art practice as well as his own theoretical work.[9] Although it is common to read that 'the impact of Burnham's work was limited',[10] Penny's account suggests that its impact was perhaps much greater among artists than among critics and historians. Indeed, the influence of *Beyond Modern Sculpture* and the important essays in *Arts* magazine and *Artforum* therefore cannot be measured in footnotes. However, a significant proportion of anglophone artists who came of age during the span of *Beyond Modern Sculpture*'s five editions, printed between 1968–78, knew about Burnham and his theories. Now, nearly half a century after its publication, Burnham's work is suitably historical, and its prescience sufficiently verifiable. As a result, his aesthetic theories are becoming much more palatable to contemporary art historians.

By 2000, it had become increasingly apparent that the exclusion and ghettoization confronting the practice and criticism of new media art and the larger historiography of art and technology required an explicit suturing strategy. In 'Art in the Information Age' (2001) I argued that by 'interpreting conceptual art and art-and-technology as reflections and constituents of broad cultural transformations during the information age' categorical distinctions can be relaxed, allowing parallels to be drawn between seemingly diverse practices, offering new insight into contemporary art.[11] Informed by Burnham's theory of systems aesthetics and his notion of software as a metaphor for art, my analysis of works by Levine, Haacke and Kosuth in 'Software' led to the conclusion that in the information age, 'meaning and value are not embedded in objects, institutions or individuals so much as they are abstracted in the production, manipulation and distribution of signs and information'. (436) Finally, I implicitly applied Burnham's systems approach to analyse the system by which art history is written. Using Haacke and Ascott as examples, I claimed that the historicization of an artist's work as conceptual art or art and technology 'says less about their work than it does about the institutional mechanisms that have created and reinforced categorical distinctions ... at the expense of identifying continuities between them'. (438)

On top of these early art-historical reappraisals of systems aesthetics, after the English publication of Niklas Luhmann's *Art as Social System* in 2000, Burnham's brilliant oddball 1960s theory gained high-powered company. A staggering number of publications addressing Burnham and his ideas were produced in the 2000s, including work by more mainstream scholars. This point is important because, as Gere has noted, a 'problem facing discourse concerning so-called new media art was how it had been contextualized and historicized ... not that there was no critical discourse, but rather that it remains the preserve of those involved, with little or no connection or engagement with outsiders.'[12] Bridging that gap, Lee embraces Burnham's theory of systems aesthetics, asserting that 'the impact of systems discourse within both the sciences and humanities is immeasurable ... its rhetoric informs and certainly facilitates a new understanding of many of the artistic practices of the 1960s.'[13]

As in 'Art in the Information Age', many of these art-historical recuperations directly confront discourses that spurned or ignored Burnham's theories. Similarly, they draw parallels between systems aesthetics and other, more authorized methods in order to identify continuities and erode categorical distinctions between the historical and current discourses of new media and mainstream contemporary art. For example, Halsall has engaged Burnham's systems aesthetics in a discourse with Luhmann, Arthur Danto, Rosalind Krauss, Nicolas Bourriaud and other writers, proposing a systems-theoretical method that draws together diverse forms of art practice and interpretative models.[14] In 'All Systems Go: Recovering Hans Haacke's Systems Art' (2008), Skrebowski took on art historian Benjamin Buchloh, using Burnham's 'Systems Aesthetics' to counter Buchloh's strict division of Haacke's work into two camps, before and after the influence of systems aesthetics: 'those earlier projects that emphasized "physiological, physical and biological processes" and the "mature" – i.e. political – works'.[15] He claims that Buchloh's antipathy toward systems aesthetics blinded him from registering Haacke's ongoing concern with systemic approaches to art that provide continuity between his biological and political works: 'Recovering the influence of Burnham's systems aesthetics on Haacke encourages us to understand his practice holistically, revealing a fundamental consistency underlying its stylistic diversity' (61). Turning Buchloh's words against him, Skrebowski argues that his position is founded on a binary opposition between nature and society: 'for Buchloh, Haacke's art cannot be political until he "transfers his interests from biological and physical systems to social systems"'.[16] Skrebowski deconstructs this mythic division and concludes:

> Systems theory offers a way to think the natural and social analogically, and Haacke's art, via his engagement with Burnham's systems aesthetics, makes use of it to do exactly that. We can now see once more that Haacke's critical artistic interventions build on an unbroken, ascending scale of systemic complexity – from organic elements, through plants, animals, and finally up to human beings. (61)

Haacke explicitly eschewed hierarchical judgements between biological and social systems. Burnham likely would agree with Skrebowsky's systemic interpretation. Its recognition of the recapitulation of fundamental orders, relations and structures at various levels of organization parallels alchemy, structuralism and kabbalah, all highly refined theories of systemic relationships that fascinated him. Within the emerging historiography of systems aesthetics, Skrebowski's reappraisal of Haacke and his dismantling of Buchloh's position demonstrate the hermeneutic potential of the systems approach.

In 'Art After Philosophy' (1969), Joseph Kosuth stated that 'art "lives" through influencing other art, not by existing as the physical residue of an artist's ideas. The reason why different artists from the past are "brought alive" again is because some aspect of their work became "usable" by living artists'.[17] Kosuth's biological metaphor suggests that art is a quasi-living organism, an open system whose elements have relevance only when they participate in the current functioning of the organism. The same claims can be made of art-historical interpretations. Were I not so sensitive to that issue perhaps fewer words would have been dedicated to this inevitably self-promotional recitation of my own historiographical contributions. I know that by interpreting and commenting on my own ideas and inserting [and reinserting] them into a living discourse I revitalize them.

Postscript

Artforum, the journal that published 'Systems Aesthetics' in 1968, later ignored Burnham, whose name was invoked in its pages only twice between 1998 and 2007.[18] It then rediscovered Burnham in 2012, celebrating 'Systems Aesthetics' and the 'Software' exhibition. In the context of my 'strategic historiography' this renewed interest in Burnham by a prominent art journal was a double-edged sword. On one edge, such mainstream recognition vindicated years of work conducted in relative obscurity; on the other edge, *Artforum* ignored the scholarly work that initiated the process of recovering Burnham from the rubbish heap of history. Neither Caroline Jones' essay 'Systems Symptoms' nor Anne Wagner's 'Data Almanac' mention Bijvoet, Gere, Penny, Whitelaw, Halsall, Skrebowski, myself or any of the artists, curators and scholars (many of whom are connected with new media art) who have contributed to this project since the mid 1990s. It is as though *Artforum* rediscovered Burnham's work on its own, effectively crediting itself for this important recuperation, without acknowledging the prior scholarship, including this historiographical study of that literature. Furthermore, in the same issue of *Artforum*, Claire Bishop's essay 'Digital Divide' limited its discussion to 'the mainstream art world' and dismissed the 'sphere of "new media" art' as a 'specialized field of its own'. Thus, even as Bishop acknowledges the presence of new media art, she condones an account of contemporary art that brackets it out of the conversation, thereby reifying the gap between mainstream contemporary art and 'new media art' that she ostensibly seeks to address. A further analysis following the approaches of Pierre Bourdieu and Niklas Luhmann offers useful insights into the systemic nature of these events.

In 'The Field of Cultural Production ...' (1983) Bourdieu notes that 'the literary or artistic field is at all times the site of a struggle between ... those who dominate the field economically and politically [in this case *Artforum* and its contributors]

... and [those] who are least endowed with specific capital [scholars of new media art, and] tend to identify with a degree of independence from the economy...'[19] (321) The art journal chose Jones and Wagner – distinguished senior art historians – due to 'the value which the specific capital of [those] writers ... represents for the dominant fractions ... in the struggle to conserve the established order...' (322) The journal's failure to cite the work of the writers associated with new media art who have done the heavy lifting on re-evaluating Burnham's work constitutes an act of rhetorical violence by omission, with several effects: 1) it systematically strips originality and authenticity from that which is excluded from the journal's pages; 2) it usurps a field of scholarship and establishes the journal's dominance over that field in its own terms; 3) it shields mainstream contemporary art discourses from interlopers that potentially threaten the status quo; and 4) simultaneously reifies the journal's position of dominance as the arbiter of those discourses. As Bourdieu observes, 'the fundamental stake in literary struggles is the monopoly of literary legitimacy, i.e., *inter alia*, the monopoly of the power to say with authority who is authorized to call himself a writer.' (323) In other words, the journal wields 'the power to consecrate [certain] producers' at the expense of others. 'One of the difficulties of orthodox defence against heretical transformation of the field by a redefinition of the tacit or explicit terms of entry is the fact that polemics imply a form of recognition; an adversary whom one would prefer to destroy by ignoring him cannot be combated without consecrating him'. (323) Much better to ignore them, bracket them out, leave them invisible ...

By contrast, in *Art as Social System* (2000) Luhmann argues that the robustness of a 'complex system' can be demonstrated by how it is capable of "processing a greater amount of irritation internally, that is, it can increase its own complexity more rapidly" (158). Following this approach, the present collection of texts aims to demonstrate the ability of art as an autonomous, autopoietic system to accommodate competing discourses that might otherwise undermine its operative closure.

1 Luke Skrebowski, 'All Systems Go: Recovering Hans Haacke's Systems Art', *Grey Room*, no. 30 (Winter, 2008) 54–83.

2 Charlie Gere, *Art, Time and Technology* (Oxford: Berg, 2006) 117, 120.

3 Roy Ascott, *Telematic Embrace: Visionary Theories of Art, Technology and Consciousness*, ed. Edward A. Shanken (Berkeley and Los Angeles: University of California Press, 2003). Man-machine symbiosis: 129; remote collaboration: 146; quote: 157.

4 Marga Bijvoet, *Art as Inquiry: Toward New Collaborations Between Art, Science and Technology* (Bern: Verlag Peter Lang, 1997).

5 Jack Burnham, *Beyond Modern Sculpture* (New York: George Braziller Inc., 1968); 'Systems

Aesthetics', *Artforum*, vol. 7, no. 1 (September 1968) 30–35.

6 Edward A. Shanken, 'Gemini Rising, Moon in Apollo: Attitudes on the Relationship Between Art and Technology in the US, 1966–71', in A. Nereim, ed., *ISEA97* (Proceedings of the 8th International Symposium on Electronic Art) (Chicago: ISEA97, 1998).

7 Edward A. Shanken, 'The House That Jack Built: Jack Burnham's Concept of Software as a Metaphor for Art' (1998), *Leonardo Electronic Almanac*, vol. 6, no. 10 (November 2008).

8 Mitchell Whitelaw, '1968/1998: Rethinking a Systems Aesthetic', *ANAT Newsletter*, no. 33 (May 1998).

9 Simon Penny, 'Systems Aesthetics and Cyborg Art: The Legacy of Jack Burnham', *Sculpture*, vol. 18, no. 1 (January/February 1999). Accessed 15 August 2009 at http://creative.canberra.edu.au/mitchell/ and http://ace.uci.edu/penny

10 Matthew Rampley, 'Systems Aesthetics: Burnham and Others', *Vector* e-zine, no. 12 (January 2005).

11 Edward A. Shanken, 'Art in the Information Age: Technology and Conceptual Art', *SIGGRAPH 2001 Electronic Art and Animation Catalog* (New York: ACM SIGGRAPH, 2001) 8–15. Quoted from expanded reprint in *Leonardo*, vol. 35, no. 3 (August 2002) 433.

12 Francis Halsall, *Systems of Art: Art, History and Systems Theory* (Bern: Verlag Peter Lang, 2008).

13 Charlie Gere, 'New Media Art', *The Art Book*, vol. 12, no. 2 (2005) 6-8. Paraphrased and quoted in Halsall, Systems of Art, op. cit., 121–2.

14 Pamela M. Lee, *Chronophobia: On Time in the Art of the 1960s* (Cambridge Massachusetts: The MIT Press, 2004) 66–7.

15 Benjamin H.D. Buchloh, 'Hans Haacke: Memory and Instrumental Reason', in *Neo-Avantgarde and Culture Industry* (Cambridge, Massachusetts: The MIT Press, 2000) 205, 212, 215; quoted in Skrebowski, op. cit., 59.

16 Benjamin H.D. Buchloh, 'Hans Haacke: The Entwinement of Myth and Enlightenment', in Hans Haacke: 'Obra Social' (Barcelona: Fundació Antoni Tàpies, 1995) 49; quoted in Skrebowski, op. cit., 61.

17 Joseph Kosuth, 'Art After Philosophy', *Studio International* (1969) (http://www.ubu.com/papers/kosuth_philosophy.html), accessed 15 August 2009.

18 John A. Tyson, 'The Afterlives of "Systems Aesthetics"', Paper delivered at Critical Information conference, School of Visual Arts, New York (8 December 2013). (http://criticalinformationsva.com/wp-content/uploads/2013/11/Tyson_John.pdf)

19 Pierre Bourdieu, 'The Field of Cultural Production, or The Economic World Reversed', *Poetics*, vol. 12, no. 4–5 (November 1983) 311–24.

Edward A. Shanken, abridged and revised extracts from 'Reprogramming Systems Aesthetics', in Simon Penny, et al., eds, *Proceedings of the Digital Arts and Culture Conference* (University of California, Irvine, 2009) (Berkeley and Los Angeles: University of California Press, 2010); reprinted in *Relive: Media Art Histories*, ed Sean Cubitt and Paul Thomas (Cambridge, Massachusetts: The MIT Press, 2013) 83–96. Postscript for this volume, 2014.

Francis Halsall
Systems of Art//2008

By the end of the 1960s the interest in the application of systems thinking by the military-industrial complex began to filter into cultural life. Between the years 1966 and 1972 there were a number of important exhibitions and publications that took the idea of systematicity as their central organizing principle, with titles such as *Systems*; *Information*; *Software* and *Radical Software*. [...] The journal *Radical Software* [...] explored the intersections between technological systems and art. Eleven editions were published by The Raindance Corporation [...] formed in New York in 1969 by the artist Frank Gillette [as] 'an alternative media think tank: a source of ideas, publications, videotapes and energy providing a theoretical basis for implementing communication tools in the project of social change'.[1] [...]

As Marga Bijvoet noted [...]:

> Cybernetics and systems notions with their accompanying vocabularies were mainly applied to the possibilities of new media systems, such as video, cable, satellite, etc. Words like system, feedback, information, software, parameter, entropy and negentropy, process, pattern became the principal vocabulary in the writers' argumentations.[2]

In the midst of the curatorial and publishing activity was Jack Burnham's own exhibition 'Software' (The Jewish Museum, New York, 1970). [...] [T]he uniqueness of the show lay in its attempt to express thoroughly Burnham's concept of systems aesthetics:

> In contrast to the numerous art and technology exhibitions which took place between 1966–1972, and which focused on the aesthetic applications of technological apparatus, 'Software' was predicated on the ideas of 'software' and 'information technology' as metaphors for art. He conceived of 'software' as parallel to the aesthetic principles, concepts or programs that underlie the formal embodiment of the actual art objects, which in turn parallel 'hardware'.[3]

The show contained [...] innovative exhibits exploring ideas of systematicity, interactivity; the use of new technological systems in art making and the shift from singular art objects to systems of art. [...] Many works were conceived of as completely interactive with full visitor participation. [...]

As Burnham explained, 'Software' was

> … an attempt to produce aesthetic sensations without the intervening 'object'; in fact, to exacerbate the conflict or sense of aesthetic tension by placing works in mundane, non-art formats.[4]

As the rationale and exhibits in the show demonstrate, he was able to do this by invoking the concept of 'software' as both a central dynamic and metaphor for an interactive art practice. To thus understand art as software is to invoke the paradigm of the coding of a computer program which is not copied in different hardware, but is rather given other manifestations. Les Levine, for example, saw his contribution to 'Software' in these specific terms, and wrote the following description of *Systems Burn-off X Residual Software* in the 'Software' catalogue: 'In many cases an object is of much less value than the software concerning object. The object is the end of a system. *The software is an open continuing system.*'[5]

More recently, Lev Manovich argued that the application of the metaphor of 'software' can be extended beyond computers to a definition of art since the advent of modernism in general:

> To summarize: from the new vision, new typography and new architecture of the 1920s we move to new media of the 1990s; from 'a man with a movie camera' to a user with a search engine, image analysis and visualization programs; from cinema, the technology of seeing, to a computer, the technology of memory; from defamiliarization to information design. *In short, the avant-garde becomes software.* This statement should be understood in two ways. On the one hand, software codifies and naturalizes the techniques of the old avant-garde. On the other hand, software's new techniques of working with media represent the new avant-garde of the meta-media society.[6]

Manovich, like Burnham, was looking for an adequate vocabulary to describe a variety of diverse artistic practices. Both recognized the potential in systems theory to reframe a discussion of the avant-garde. For Burnham in particular this entailed a formulation of an aesthetic theory based upon the central paradigm of the possibility for a system to be conceived as a medium for art. […]

To understand art as software is to understand it in terms of codes and information rather than in material or medium-specific terms. Burnham proclaimed that systems aesthetics necessitated the dissolution of the material specificity of traditional artistic mediums, so that the traditional 'objet d'art' would eventually be replaced by 'aesthetic systems'.

> [The] cultural obsession with the art object is slowly disappearing and being replaced by what might be called 'systems consciousness'. Actually, this shifts from the direct shaping of matter to a concern for organizing quantities of energy and information.[7]

In his advocacy of 'the organizing quantities of energy and information' for a new type of art-making, Burnham advanced an art that was both ontologically unstable and interactive. He thus specifically linked a post-formalist artistic attitude – prevalent in the art world at the time, which was also interested in exploring de-ontologized and interactive art – to the contemporary discourse of systems theory. In 1968 he wrote:

> The post-formalist sensibility naturally responds to stimuli both within and outside the proposed art format ... [but] the term systems aesthetic seems to encompass the present situation more fully.[8]

In doing so, Burnham articulated in systems-theoretical terms the emerging historical situation in art practice toward the end of the 1960s. [...] Systems aesthetics can also operate as what Peter Osborne called a 'retrospective critical discourse', which 'does not need to discover its terms literally or empirically within the discourse of the period under discussion'.[9] This means that systems aesthetics can also be identified as a function of the discursive system from which it is observed and constituted; it thus can be integrated into a coherent historical and sociological narrative.[10] [...]

It is my argument here that Burnham's systems aesthetics is compatible with a variety of art-historical descriptions and can therefore be employed as part of a retrospective critical discourse of systems aesthetics. Thus conceived, systems aesthetics allows for an expanded field of practice, implying a shift from singular art objects to the use of systems as artistic mediums. These descriptions include the 'dematerialized' art object (Lucy Lippard), 'intermedia' (Dick Higgins) and the 'post-medium condition' (Rosalind Krauss).

Lucy Lippard's definition of the 'dematerialized' art object highlighted what was at stake when the anti-modernist aesthetic led to a dismantling of the modernist art object. [...] Such work can be understood as exploring an aesthetics of systems, and in doing so thus functioned by investigating the ways in which it was embedded in various networks of display, representation, meaning and control.

Dick Higgins used the term 'intermedia' in 1966 to explain [how] 'much of the best work being produced today seems to fall between media'. [...] Given its plurality, intermedia art is resistant to an account of it in terms of the material specificity required of it by a limited account of modernism. Thus the term 'intermedia art' is, like Lippard's phrase, compatible with Burnham's account of a practice that attempted to make art without the production of unique art objects. [...]

Krauss's notion of the post-medium condition takes account of a situation in the 1960s which was also described by Lippard, Higgins and crucially, Burnham. This was a situation when artists began to form a critique of modernist medium-specificity. The systems aesthetic thus provides the opportunity to reinscribe materially diverse practices (such as minimalism, conceptualism and new media art) within the single medium of system. In doing so the 'retrospective critical discourse' of systems theory provides the vocabulary to map different artistic strategies within a single rubric. One may thus discuss historical examples as contiguous with more recent ones. [...]

Recent curatorial interest in systems has demonstrated both its effectiveness as a descriptive paradigm and its relevance to contemporary practice. For example, in 'Open Systems: Rethinking Art *c.* 1970' (Tate Modern, 2005), curator Donna De Salvo [...] demonstrated that as an organizing principle system was flexible and suggestive enough for 'rethinking' historical practice whilst also providing a focus that is effective, rigorous and engaging (and popular). The concept of system thus provided an opportunity to re-evaluate the chosen work in a broader historical and aesthetic context in which the often unhelpfully narrow titles of minimalism, conceptualism and so forth could be sidestepped in favour of critical descriptions grounded upon a shared interest in systems of display, representation, meaning and control. [...] [T]he exhibition uncovered interesting connections between artists who produced work as visually diverse as Robert Smithson, Carl Andre, Lygia Clark, Andy Warhol and Bruce Nauman. To reflect this systemic connection [...] all works on display were 'linked by their use of a generative or repetitive system as a way of redefining the work of art, the self and the nature of representation'.[11] [...] De Salvo's claim on the persisting relevance of systems art is supported by three instances of contemporary curatorial and artistic interest in systems.

First, several shows between 2004 and 2006 concentrated on a reappraisal of art from the late 1960s and early 1970s, including high profile retrospectives of Donald Judd [in Europe], Dan Flavin and Robert Smithson [in the US]. This was followed by contemporary installations by artists who had been active since the 1960s, including Bruce Nauman's *Raw Materials* (2004–5, Tate Modern) and Richard Serra's 2006–7 installations at the Guggenheim in Bilbao and New York. All this activity reinforced the cues that contemporary artists (for example, Liam Gillick, Carsten Höller and Mark Dion) were clearly taking from conceptualism, minimalism and institutional critique, and thus demonstrated how systems could be employed to map an expanded field of practice from historically different times.

Second, such a mapping would draw direct connections with other artists who have specifically investigated how their own bodies are situated within various social, physical and psychic systems. In [the artist Stelarc's] bodily modifications he subordinates his physical body to cybernetic and technological systems. For example, in *Ping-Body* (1996), his body was controlled by prosthetic extensions remotely controlled by users over an internet connection. Brian Eno's *77 Million Paintings* (2006) used self-generating and developmental algorithmic systems to create a sequence of evolving patterns or 'paintings' displayed on monitors and television screens. These examples demonstrate the diversity of art practices whose predominant mediums can be observed as systems.

Third, renewed contemporary interest in self-critical practice that expands its scope beyond the limits of a singular artwork into the social systems within which it is placed has been the specific tenet of French curator Nicolas Bourriaud's [...] concept of 'relational aesthetics' since the start of the 1990s (again, the examples of Liam Gillick, Carsten Höller and Mark Dion are appropriate). [This] has pushed for a move into 'the realm of human interaction and its social context, rather than the assertion of an independent and private symbolic space'.[12] Such work arguably takes its cues from the rigorous probing of the status of art and its institutions that is characteristic of art that engages in systems aesthetics (for example, Burnham and Haacke's work from the late 1960s and early 1970s). Bourriaud's *Relational Aesthetics* argues that art institutions might become [...] part of the social systems that create new venues for relations among different practices and viewers; they are both 'frame' for the work and part of its medium. Relational aesthetics is thus compatible with systems aesthetics in so far as it radically reconceives the purposes and effects of art practice, and thus puts into question common notions of the nature of art objects. This reconceived understanding locates art in a system of relationships between art and its environment, its viewers and art discourse.

I close with a further example of how the notion of system might be usefully employed by contemporary art historians. This is in providing an account of so-called 'new media art'. New media art has had a problematic reception in art-historical terms. As Charlie Gere observed, it raises the question as to what critical discourse is supposed to deal with it: 'If new media art wishes to be taken seriously then it is necessary to start to develop appropriately robust and convincing means by which it can be examined critically.' [...] [T]he point was 'not that there is no critical discourse, but rather that it remains the preserve of those involved, with little or no connection or engagement with outsiders.'[13]

Gere's claim is one that art historians should take very seriously. However, this requires [overcoming] the traditional art-historical preoccupation with specificity of media and [how this problem may be reconciled] with the proliferation of differing media employed in new media art. W.J.T. Mitchell diagnosed the problem in the following terms:

> In the field of art history, with its obsessive concern for the materiality and 'specificity' of media, the supposedly 'dematerialized' realm of virtual and digital media, as well as the whole sphere of mass media, are commonly seen either as beyond the pale or as a threatening invader, gathering at the gates of the aesthetic and artistic citadel.'[14]

Systems aesthetics and the vocabularies of systems theory provide, I argue, the basis for such an 'appropriately robust and convincing' theory of new media art. They do so by expanding the discourse on new media beyond a discussion of a narrow set of art practices corresponding to a limited set of media into a discussion about systems art more generally. Systems theory and systems aesthetics thus employ the idea of system as medium to inscribe a coherency into what would otherwise seem to be utterly disparate works. This opens these works up to art-historical analysis and provides continuity with historical precedents which may, in the first instance, appear materially incomparable. [...]

1 [footnote 20 in source] Davidson Gigliotti, 'A Brief History of RainDance' (2003) on the website for *Radical Software* (http://www.radicalsoftware.org/e/index.html) (10/09/2007)

2 [22] Marga Bijvoet, *Art as Inquiry* (Bern: Peter Lang, 1997) 75.

3 [23] See Edward A. Shanken, 'The House that Jack Built' (1998), *Leonardo Electronic Almanac*, vol. 6, no. 10.

4 [26] Jack Burnham cited in ibid., from personal correspondence with the author.

5 [27] Quoted from N. Wardrip-Fruin and N. Montfort, eds, *The New Media Reader* (Cambridge, Massachusetts: The MIT Press, 2003) 255; emphasis added.

6 [28] Lev Manovich, 'Avant-garde as Software', *Artnodes* (http://www.uoc.edu/artnodes/eng/art/manovich1002/manovich1002.html), accessed 22 March 2006; emphasis added.

7 [30] Jack Burnham, *Beyond Modern Sculpture* (New York: George Braziller Inc., 1968) 369.

8 [31] Jack Burnham, 'Systems Aesthetics', *Artforum* (September 1968).

9 [32] Peter Osborne, Presentation for symposium on 'Open Systems: Rethinking Art *c*. 1970', Tate Modern, London (16–17 September 2005).

10 [33] And acknowledging this means recognizing the relationship between the discursive position of systems theory and the historical phenomenon that it both observes and constitutes by virtue of that observation. [...]

11 [40] Donna De Salvo, introductory essay, *Open Systems* (London: Tate Publishing, 2005).

12 [41] Nicolas Bourriaud, *Relational Aesthetics* (1989); English edition (Dijon: Les Presses du Réel, 2002) 14.

13 [43] Charlie Gere, 'New Media Art', *The Art Book*, vol. 12, no. 2 (2005) 6–8.

14 [44] W.J.T. Mitchell, *What Do Pictures Want?* (Chicago: University of Chicago Press, 2005) 205.

Francis Halsall, extracts from *Systems of Art: Art, History and Systems Theory* (Bern: Peter Lang, 2008) 106–23.

Caroline A. Jones
Systems Symptoms: Jack Burnham's 'Systems Aesthetics'//2011

A brief meteor, Jack Burnham blazed forth in September 1968 declaring 'Systems Aesthetics' to be the pre-eminent mode of contemporary art-making. How could this sculpture teacher from the Midwest have gotten it so right? In the ensuing decades, what Burnham called 'systems' came to define some of the most significant cultural developments of our time, even if this genealogy has been obscured. Undeclared as such, systems thinking now suffuses the art world as we know it. Consider: The work of such different artists as Olafur Eliasson, Andrea Fraser, Damien Hirst, Seth Price and Tino Sehgal, as well as platforms such as Triple Canopy, would be hard to understand without recourse to some concept of art as a disparate, sprawling, yet rule-bound system within which artists must strategically acknowledge dealers, viewers, performers, participants, buyers, fabricators, curators, programmers, institutions, infrastructures, the environment, magazines such as *Artforum*, algorithms, and other constellations around and within the ever more expansive work of art.

Burnham called it. And *Artforum*, under editor John Coplans, continued to push it, with Lawrence Alloway's 'Network: The Art World Described as a System' appearing in 1972. But Burnham's thesis has become unaccountably obscure. One factor was the revulsion that soon confronted the very term systems and the source of its discourse – Burnham's first footnote refers to the Rand Corporation's 1964 publication *Analysis for Military Decisions*, and the obvious entanglement of systems theory with the military-industrial complex was a fatal attribute in the eyes of his largely leftist audience during the Vietnam era. By 1980, Burnham himself was unwilling to continue plumping for this unlikely constellation of war games and process art, castigating his early techno-utopianism as a 'panacea that failed'.[1] But the ultimate answer may be that systems didn't fail: They simply wormed their way into the art world like a weakened virus. Contemporaneity is built on Burnham-like 'systems'; their code laces our cultural genome. Exploring the evolution from Burnham's vision of a 'systems aesthetics' at the core of art-world discourse to our current reality of systemic practices not only opens up alternative historical perspectives on the intervening decades but suggests a fresh view of much contemporary art. The systems virus is still here – but it has mutated from Burnham's engineering and industrial setups into the social and electronic protocols that govern our world. Burnham bemoaned that the systems aesthetic 'has no critical vocabulary so necessary for its defence'; what we got instead was a critical vocabulary utterly defended from its own sources in systems thinking.

Reading the original 1968 essay today is an exercise in reverse teleology. It reveals Burnham as an ambitious player in the high-stakes gamble over the legacy of modernism initiated in the 1950s by Clement Greenberg and carried on by Michael Fried. [...] Burnham's systems outlined a counter-trajectory that laid claim to the entire future of contemporary art.[2] Systems artists were not only 'post-formalist' (which Burnham argued explicitly) but 'post-Minimalist' (as *Artforum* editor Robert Pincus-Witten would soon declare). In 'Systems Aesthetics' [...] Burnham recognized physical 'systems' that enmeshed the viewer in a range of embodied, conceptual and planetary entanglements that art would both instantiate and reveal.

Despite his own debt to phenomenology, Fried's theories of vanguard modernism (like Greenberg's) famously demanded a strict separation of art from the world. [...] Ambulatory environments, let alone the 'processing' of the viewer through some durational engagement, were part of an abjured theatricality. By contrast, Burnham flatly rejected the object and the segregation it implied, engaging systems that provoked phenomena and allowed the systems artist to reduce 'the technical and psychical distance between his artistic output and the productive means of society'. [...] Burnham argued that 'in an advanced technological culture the most important artist best succeeds by liquidating his position as artist vis-à-vis society'.[3] [...] [T]he application of a 'systems aesthetic' could render form secondary – and thus open up art practice to a wide range of urgent issues including 'such concerns as maintaining the biological livability of the Earth, producing more accurate models of social interaction, understanding the growing symbiosis in man-machine relationships, establishing priorities for the usage and conservation of natural resources, and defining alternate patterns of education, productivity and leisure'.[4]

It seems clear that part of Burnham's agenda was to arrogate for artists goals that could rival the systems being developed elsewhere (from Landsat to IBM). In 1968, such systematic ambitions were epitomized by the work of Dan Flavin, Les Levine and, even more consistently, Hans Haacke, on whom Burnham had just published a first monograph.[5] Burnham's support for Haacke, and Haacke's admiration for Burnham, vectored them both to MIT – arguably the narthex of all things system in 1967–68.

Burnham must have finished the Artforum piece just before moving to MIT, where he joined the first cohort of fellows at Gyorgy Kepes' Center for Advanced Visual Studies and pitched a course on 'Systems and Art' he had developed with a systems engineer, Gustave Rath. Kepes was already actively engaged with systems theory at the time, having solicited texts from biologist Ludwig von Bertalanffy (whose writings Burnham footnoted in the 'Systems Aesthetics' essay) and cybernetics theorist Norbert Wiener.[6] But where Kepes would see his job as ameliorating the alien cybernetic world of 'command, control and communication', Burnham subscribed to total engagement. His 'Systems and Art' course and the aesthetics essay were staging grounds for techno-sociality in itself, not merely 'a novel way of rearranging surfaces and spaces' but 'fundamentally concerned with the implementation of the art impulse in an advanced technological society'.[7] This grand view was refined in a second *Artforum* article, published in September 1969: 'Real Time Systems'. Here Burnham compared artists to 'programs and subroutines', their work 'an archaic information processing system' we need in order to 'prepare new codes and analyse data'.[8] Art is not special in this regard, exhibiting the same protocols as other knowledge-producing activities, but it is better at revealing the constructedness of consciousness. Segueing to the computer-controlled real-time systems coordinating contemporary economic and military domains, Burnham acknowledged the 'instinctive antipathy' that most humanists have toward 'these immensely complex computer systems. Their Orwellian overtones far overshadow their conceivable use as artists' tools. But practically, it is imperative that artists do understand them – both technically and philosophically.'[9]

The radical consequences of this imperative were revealed by the legendary cancellation of Hans Haacke's planned retrospective at the Solomon R. Guggenheim Museum in the spring of 1971. Exemplifying the 'liquidation' of the artist in a way Burnham had never anticipated, the Guggenheim cancellation demonstrated that systems art had merged with society's systems and was thereby subject to their political, legal and economic forces. The cancellation was thus both the apotheosis of systems and the moment after which it would become difficult to say its name. In this light, it is moving to read Burnham's response to the exhibition that never happened, writing in June 1971 of Haacke's plan to work with 'physical, biological and social systems', the totality revealing the artist's commitment to 'the interconnectedness of all systems' and an art that could interrogate 'the way the world functions on its most essential levels'. Given the cancellation, Burnham was forced to conclude that the Guggenheim had decisively altered Haacke's 'systems art', which could therefore be understood as drawing closer 'in *its semiotic structure* ... to the *ritual drama* (where the artist's *premises* are *recapitulated* in *everyday life*) and *away* from the *plastic arts*'.[10] (The emphases in both instances are Burnham's.) Never more prescient than here in his anticipation of the shift from techno-mechanical systems to social protocols, Burnham nonetheless lost the capacity to shape artistic discourse, or to advocate for 'systems' as a concept. Henceforth 'systems', with their uncontrollable boundaries, would be abandoned for a more tractable art-world discourse of 'institutional critique', one that *Artforum* authors and editors would be active in promoting. We might consider whether institutional critique is, indeed, our 'ritual drama', with the secret life of systems that undergirds it emerging later, and elsewhere, in unacknowledged ways.

The Guggenheim cancellation may have cemented a shift to the social that was already under way in Burnham's thinking. [...] In the introduction to his own anthology of systems essays, Burnham was surprisingly personal about the function they had served: 'Ultimately systems theory may be another attempt by science to resist the emotional pain and ambiguity that remain an unavoidable aspect of life'.[11]

But systems' virions survived. By the 1990s, what I would call systemic artworks had emerged with a vengeance. Just as Burnham had recognized an array of physical and material entanglements in work otherwise described as conceptual or dematerialized, systemic artworks dialectically reject or critically torque the virtual ideologies of the Internet to materialize the links that join archival, research-driven, process-oriented, labour-intensive, recursive, informational, social and communicational aspects of art. In 2012, Burnham's concerns about everything from the consumption of natural resources to the implementation of machine technology seem tailor-made for the contemporary art world. Take Documenta 13 this past summer, where such systematics were witnessed in the crowd of artists offering soil currency (Claire Pentecost), time banks (Julieta Aranda and Anton Vidokle), off-limits apiaries (Pierre Huyghe), or machine-propelled air currents (Ryan Gander). And Burnham's words to Kepes from November 1967 are entirely in line with the Arup-infused logic of much contemporary art and architecture: 'What I propose is not that the artist become an engineer, but that for some phases of his problem-solving it would be advisable for him to think like an engineer [managing] input-output exchanges of materials, energy and information.'[12] If Burnham turned from the systems he prophesied, we would find it impossible to do so. We cannot turn from them because they are turning within us, the dynamic engine of our imbrication in many aspects of lived reality – the art world, but also the economic, legal, ecological and political worlds we navigate and that seem to implode or explode daily. Whether or not we want to see or name them, systems are us.

1 Jack Burnham, 'Art and Technology: The Panacea That Failed', in *Myths of Information: Technology and Postindustrial Culture*, ed. Kathleen Woodward (Madison, Wisconsin: Coda, 1980) 200–215.

2 Pamela M. Lee, *Chronophobia* (Cambridge, Massachusetts: The MIT Press, 2004) 74.

3 [footnote 5 in source] Jack Burnham, 'Systems Aesthetics', *Artforum*, vol. 7, no. 1 (September 1968) 31.

4 [6] Ibid.

5 [7] Jack Burnham, *Hans Haacke Wind and Water Sculpture: supplement to Tri-Quarterly* (Evanston, Illinois: Northwestern University Press, 1967).

6 [8] Wiener gave permission for a reprint of 'Pure Patterns in a Natural World' for Gyorgy Kepes' 1956 collection *The New Landscape in Art and Science*, and Ludwig von Bertalanffy contributed 'The Tree of Knowledge' to Kepes' 1966 *Sign, Image, Symbol.*

7 [9] Burnham, 'Systems Aesthetics', op. cit., 35.

8 [10] Jack Burnham, 'Real Time Systems', *Artforum*, no. 8, no. 1 (September 1969) 49.

9 [11] Ibid., 51. This assertion was accompanied by a picture of Haacke's *Chickens Hatching* (1969).

10 [12] Burnham, 'Hans Haacke's Cancelled Show at the Guggenheim', *Artforum*, vol. 9, no. 10 (June 1971), reprinted in Amy Baker Sandback, *Looking Critically* (Ann Arbor: UMI Research Press, 1984) 105–9.

11 [13] Jack Burnham, 'Introduction', in *Great Western Salt Works: Essays on the Meaning of Post-Formalist Art* (New York: George Braziller Inc., 1974) 11.

12 [14] This quote is from the course Burnham had developed with engineer Gustave Rath and proposed to teach at MIT. Annotated in Burnham's handwriting 'for G. Kepes', the typescript is titled 'Some thoughts on systems methodology applied to art'. Burnham file, Center for Advanced Visual Studies archive, MIT, provisionally dated with the letter that refers to it, November 1967, ts p. 1.

Caroline A. Jones, extracts from 'Systems Symptoms: Jack Burnham's "Systems Aesthetics"', *Artforum*, vol. 51, no. 1 (September 2011) 113–16.

Boris Groys
The Mimesis of Thinking//2005

The contemporary 'art system' emerged as an effect of the shift in artistic practices that took place during the 1960s and 1970s. Art is always in flux; its forms are always subjected to a historical evolution. But in the 1960s the role of the artist was subjected to a radical redefinition that has not lost its validity until now. Until the 1960s the romantic image of the artist remained fundamentally intact. The 'true' artist was understood to be a lonely creative individual following not the external rules and conventions of society, but exclusively his or her 'inner necessity', as Kandinsky famously put it [*Concerning the Spiritual in Art*, 1912]. The role of the artist was to act outside modern bureaucracies, outside the huge socio-economic machines of modern industrial production. The creative artistic act served as a paramount example of a non-alienated, liberated work. Of course, the artist, in a very acute way, experienced dependency on the capitalist art market, on the prevailing public taste, on the explicit or implicit censorship in the name of generally accepted norms and values. But the duty of the artist was seen precisely in the struggle for liberation from these external norms, values and dependencies. This struggle was regarded as possible and even necessary because the artistic creative act itself was understood as being uniquely autonomous, internally free. Art had to manifest this inner freedom openly to be recognized as 'true' art. But it is precisely this inner autonomy and freedom of the creative act that was questioned by the art practices of the 1960s and 1970s.

It was during this period that mass cultural imagery began to invade the whole visual field of contemporary society. An individual artist could no longer compete effectively with the commercial apparatuses of anonymous image production. In addition, emerging computer technology demonstrated the possibility of producing, processing and registering images without any direct intervention by a human producer or spectator. Some authors and artists reacted to this loss of individual control over image production with desperate protest [e.g. the Situationists]. Other artists developed a new strategy, which operated through the individual appropriation of mass-produced images [e.g. Pop]. In both cases the 'system' was seen as something purely external, as something opposed to the unique subjectivity of an individual artist. Retrospectively, it seems that the real shift was effectuated by minimal and conceptual art of the period, because in this context an individual artwork was understood as being inscribed in a certain system of image production and communication from the start. This shift was partially inspired by different linguistic theories, such as French structuralism or the Wittgensteinian concept of language games. Notwithstanding the differences in these details, all these theories interpreted an individual speech act as an application of a set of general linguistic rules. Accordingly, the advanced art of this time understood the individual act of art production as being originally regulated by a 'system', as following a certain general rule from the beginning, and as being inscribed into a certain social practice even before its product was submitted to a definite social use.

This change of attitude towards the notion of a system can easily be misunderstood as an act of capitulation vis-à-vis the apparatuses of technological progress and commercialized mass culture. But in fact this change allowed the artist to analyse and criticize the dominant regime of image production and distribution by his or her own artistic means for the first time. Indeed, if the act of art production is understood as fully autonomous and genuinely free, then the artist can only be a slave or a victim of the external systems of the art market, art institutions and so on, being completely heterogeneous in relationship to these systems. But if the creative act itself is part of a certain system and guided from the beginning by a certain set of rules, then the artist has a unique inner access to the system. And this means that the artist has a unique competence and power in dealing with this system. The integration of an individual creative act into a communicative system was interpreted by some theoreticians as a sign of the death of autonomous artistic subjectivity. But this subjectivity successfully survived its death by making the system itself the object of its inner, intimate experience. There are many ways to understand what kind of system is guiding the inner creativity of an artist. Some artists wanted to analyse the existing systems, the existing visual codes that compel the artist to use a certain vocabulary of images and to combine them according to a certain set of rules. Other artists tried to develop alternative, utopian systems of visual communication that would be able to supplant and substitute the existing image regime. There were artists such as Joseph Beuys or Lygia Clark who [from the 1960s onwards] developed their own myths, their own complicated systems of meaning production and communication. And there were the artists who played ironically with the socially accepted visual codes to subvert and deconstruct them, such as Marcel Broodthaers or Ilya Kabakov. Overall, the art of the 1960s shifted its focus from the individual creative act to a description, investigation and development of communication systems and visual codes. Accordingly, the art world as a whole began to be perceived as an 'art system'. The metaphysical loneliness of the romantic artist was substituted by strategies of participation and collaboration. The artist became a part of the art system, of the art bureaucracy. The artist's main occupation became not to create but to criticize. The paradoxical figure of a 'critical artist' that emerged in the 1960s announced an end to a long period of confrontation between the individual artist-creator and the art critic serving the 'system', a conflict that contributed substantially to the dynamic of romantic and modernist art. [...]

The typical Minimalist installation is perceived as a fragment of a formalized algorithm of reiterations and modifications. However powerful and fascinating the immediate visual impression of these installations on the viewer may be, ultimately they point to something invisible, merely conceivable, virtual. Clearly, the same set of binary oppositions, the same visual code that is manifested in the installations, can produce a potentially infinite row of new objects. That is why the viewer's imagination is stimulated to imagine the generative code, to imagine all the variations that can be generated by the code. Such an attempt, however, immediately points the viewer in the direction of the invisible set of rules on which the different variations are based. And that means nothing other than that an individual artistic decision is no longer understood as sovereign, as fully autonomous but, rather, as an individual application of the existing set of rules, as a realization of an option that is always already given. The same is true, of course, for Donald Judd's early installations as well as those of Carl Andre, where the objects mostly do not vary at all but are simply reiterated. Here the variations are reduced on the one hand to mere repetition within the framework of an installation. On the other hand, however, Judd varied materials and forms from installation to installation in an easily understandable, transparent way. That was also true of Andre's installations, which practise pure combinations of the simplest geometric forms. This is a strategy of variation that puts the artists beyond the traditional opposition between affirmation and negation, between repetition and innovation. Hence if Minimalist artists themselves repeatedly insisted on the immediate presence of their objects, de facto the most important aspect in Minimalist installations took place in the zone of the invisible, that is, between the art objects.

This view was formulated, with critical intent, in Michael Fried's famous essay 'Art and Objecthood' (1967). Fried criticized Minimalist installations for drawing the viewer's attention away from the individual artworks and into the space of the installation. According to Fried, in a Minimalist installation the presence of the space is felt more strongly than the presence of the art objects themselves. This is why Fried attributes a theatricality to Minimalist installation, which he characterizes as hostile to art. The reason for this critical assessment of Minimalist installation becomes clear when Fried writes that that which lies between the artworks, or even between the individual arts, can only be the theatre, the stage. To put it another way, for Fried that which takes place between artworks is always just another image – in this case an image of a stage. But, as I have tried to show, what happens between artworks in a Minimalist installation is not theatre but a set of rules, a formal logic, an algorithm, which may generate an image but is not in itself an image. Minimalist installation is also crucially distinct from theatre in that it can be walked on [in the case of Andre] and around. An installation does not present itself to visitors as a stage that can only be observed from a certain position, but as a space for the flâneur, for walking from one object to the other. The viewer's movement from one art object to another is guided by the same system of rules that determines the space between the individual artworks in an installation by linking those artworks by a series of reiterations and modifications.

One might say that Minimalist installation practises the mimesis of thinking. In that phrase, 'thinking' is understood to mean a step-by-step movement from one option to the other, from one variation to the other within an overarching, virtual system that incorporates and arranges all such conceivable options and variations. The goal of using art as a mimesis of thinking unites almost all the main trends in the art of the 1960s and 70s. It also subsumes many of the attempts to use language – as a supposedly direct representation of thinking in the context of art. [...] But the infinity of thinking cannot be represented by an individual image. Thinking progresses from one image to another in a systematic way without any conceivable end. Thinking is the infinite progressive movement, the infinite 'et cetera'. Therefore, it can be represented in the art context only in the form of an installation that recreates this progressive movement, even if only inside a finite, limited space. Minimalist installation represented the process of methodically, systematically organized thinking precisely by directing the visitor to a step-by-step movement from one object to the other. Here we are dealing with the same understanding of thinking that lies at the basis of computer programming. And of course, only the introduction of the computer made it possible to represent such thought processes and to formalize them to a greater extent. Minimal and conceptual art of the 1960s had, however, taken the decisive step in the direction of representing thought processes by taking pure thinking as its object and thus aestheticizing it. In this sense Minimalism is at the same time very much a megalomaniacal maximalism that wants to transcend the limits of the finite installation space. Not only does the artist subject the uniqueness of his or her artistic decisions to an abstract, infinite generative code, but the artwork itself also ceases to be a concrete, unique artwork, and instead presents itself from the outset as a fragment of a potentially infinite progression that, while it can certainly be understood, grasped and even continued at will, cannot be completely realized.

Now, however, every programme for mimesis leads to countless difficulties and paradoxes. Magritte observed that a representation of an apple is not an apple and a representation of a pipe is not a pipe. So too a representation of thinking by means of computer programs or artistic installations is not yet thinking. Human thinking is used for the purpose of individual and collective survival in the service of the survival instinct. 'Intelligent' machines and artistic installations do not think; they merely represent thinking beyond any concern about their own survival and wellbeing. The mimesis of thinking is thus in many respects confronted with the same difficulties that faced the mimesis of Nature. Above all, it faces the fundamental question of how one represents the system of all possible options in the necessarily limited space of the artistic installation. The thinking is potentially infinite. The space of the installation, by contrast, is finite. Minimalist installation lives from the tension that results from the encounter of an abstract, infinite generative code that regulates the production of art objects within the installation space and the external, contingent characteristics of this space whose size limits the code's further realization. This incursion of the contingent into the infinite progression of thinking which takes its internally unmotivated, 'irrational' limit from the external form of the installation space, is, however, merely an external symptom of the irrationality that internally affects every code and every system of thinking from the outset. If, for example, artists are asked why they chose precisely this rule of variation and not another, they can explain it either by falling back once again on their own contingent, subjective, creative decisions or by reference to a meta-system that determines the choice of the specific rule in each individual case. Such a reference to an ever-higher meta-level system, however, famously leads to insoluble logical paradoxes that in turn can be eliminated only by the contingent decision to limit the system. [...]

Boris Groys, extracts from 'The Mimesis of Thinking', in *Open Systems: Rethinking Art c. 1970*, ed. Donna De Salvo (London: Tate Publishing, 2005) 51–64.

Stephen Jones
A Cultural Systems Approach to Collaboration in Art and Technology//2005

[...] [In 'Systems Aesthetics' (1968), Jack Burnham] notes that 'The scope of a systems aesthetic presumes that problems cannot be solved by a single technical solution, but must be attacked on a multilevelled, *interdisciplinary* basis.'[1] (my emphasis) and it is here that the interaction between technologist and artist enters consideration. Architects, engineers and electronics technicians may all become involved in the production of a work [...], widening the network of ideas and influences that go into the production, and making the production more of a process, though Burnham does suggest that the artist may actually assume some of these functions.

Burnham recognized the role of the network of interacting individuals within a system and the functions of feedback and adaptation processes that go into developing an artwork using contemporary technologies, but he didn't explore the further consequences, which are that the interactions among those individuals are sequential, taking place over time, and that the relational ascendancies also vary over this time. In more general terms, a system is a network of nodes having disparate relations which change over time, so that at one interval one node is a source and at another a different node becomes the source, giving rise to opportunities for feedback relations to have a range of impacts both within the making of an artwork and within the greater society.

To a certain extent Burnham's analysis is incomplete because it fails to account for the potentially reciprocal interaction between the artist and the technologist. While his concern was with the role of cybernetics in the making of responsive artworks, the general thesis here is that this cybernetic activity is also what takes place between individuals, and between individuals and their institutions, in the processes of collaboration. It is this wider class of cybernetic coupling that provides the basis for my view of the relations between artists and their technical support and reciprocally the role of the artist in industrial and scientific institutions, with or without a collaboration in place. The technology makes available and the user makes demands, each feeding the other in reciprocity. There is a continual process of feedbacks (a conversation, so to speak) between the demands of the user – artist or scientist – and the engineer, the technologist. It goes something like this. Within some context the engineer develops the existing capabilities of a technology. These capabilities may stimulate the artist to utilize that technology for some process which suits their context and intentions, but the technology will be, almost necessarily, inadequate to the artist's intentions. It is here that the potential for collaboration appears. Even if it does not actually produce a collaboration, the needs of the artist can stimulate an engineer to extend the technology in some way thus extending the possibilities of its use, and thereby extending the range of the works that the artist might produce with that technology. Thus technology and art can *co-evolve* in a configuration of mutual interdependence driven by the feedback each supplies to the other, which is a cybernetic process, whether there is an active collaboration taking place or not.

While Burnham points to the value of collaboration (if only to keep the artwork running) and that it is a cybernetic system showing feedback distributed interaction, his model remains incomplete in that it doesn't elucidate the process by which collaboration evolves as a problem-solving mechanism. In order to reach towards this necessary extra layer in the interactions that constitute collaboration it will be useful to introduce Deleuze and Guattari's 'machinic phylum', through which they provide us with a view of the dynamics or the motive forces in the cybernetic system.

In their 'Balance Sheet Program for Desiring-Machines'[2] they consider our whole social process under the general rubric of the 'desiring machine' as an aspect of their 'machinic phylum'. They invoke a cyclical interaction mechanism that is cybernetic and of a greater spread of function within society. (And what follows is my extractive/interpolative reading of their article.) The machinic phylum must take hold of a tool so that

> [The person] and the tool *become* or *already are* distinct components of a machine in relation to an effectively engineering agency (*une instance effectivement machinisante*). And we believe moreover that there are always machines that precede tools, always phyla that determine at a given moment which tools, which [people] will enter as machine components in the social system being considered.'[3]

The machinic phylum is seen not as the tools and machines that we *use* but as a *dynamic network of technologies and people*, a *social machine* that functions at that higher (societal) level. It is an organization of functional nodes (people and institutions) in a social, collective network wherein the flow of energy and information produces the organization of the machine (system) and drives its evolution. This social machine functions through communication and interaction, it is recurrent (i.e. a feedback circuit), utilizing 'the probability of a less-probable' (i.e. it produces order or new states of lower entropy, which is a definition of information); not 'acting through the functional synthesis of a whole' but 'through real distinctions in an ensemble' (as through the *production* of information by the system). In many ways it represents the linkages between people that make up a society. The motivational forces that flow through these linkages, producing the dynamics of the system, are the forces that Deleuze and Guattari gather under the rubric of 'desire'. [...]

Artists could almost be thought of as paradigmatic desiring-machines but for the fact that the desiring-machine exists at the more interactive social rather than individual level. Artists are often particularly difficult to pin down, artistic creativity being quite different from the more directed activity of the engineer. As Deleuze and Guattari note: 'What defines desiring machines is precisely their capacity for an unlimited number of connections, in every sense and in all directions.'[4]

They are or become rhizomatic, proliferating in the world as sequentially coupled interactions having impacts, in varied ways, on themselves and on each other as systems in process with other surrounding systems. The desiring machine is the ensemble of individuals and 'fixed' entities (tools and machines in our usual way of speaking) the components of a constantly *inter*-looping collection of relations among components driven by our interests and desires and the tools' offerings. [...] Desire, generosity, multiple idiosyncratic behaviours drive the process, draw things in – building the desiring machine. Components mutate, producing a radical break as inventions. New approaches and discoveries energize the phylum. These breaks are step-functions in its processes (in its local and evolutionary time scales) and it shows a punctuated evolution – flow-break-flow. It self-organizes (as autopoeisis) as 'a collective *full body*, the engineering agency on which the machine installs its connections and effects its ruptures'.[5] Thus the collaborative process can be seen as this 'machine' in itself, a machine that functions through multi-layered feedback processes operating between the individuals who are the 'nodes' in the 'network' that is the organization of the machine. [...]

Within the machine we are engaged in construction – we are engaged with the world in the process of its and our construction. To draw the process apart into components is ultimately to mislead, because it is the one process of these components tightly coupled that is the world and our becoming in it. [...]

The system of a collaboration in process entails layers of interaction and feedback and they are the object of interest here. It is the cybernetic processes in which the interactions within this structurally coupled system consist that bring forth its evolution. The concept of interaction that I am invoking here, in which the internal processes of the system (artist + technologist + the devices they produce and use) act to produce and reproduce its components, setting up a sustained existence for those components in the face of environmental perturbation, leads one to the concept of Autopoiesis. [...]

[M]any of the characteristics of autopoiesis assist greatly in understanding the development of the 'components' (the term has a particular technical function) that would be operating in the social environments which produce the types of artefacts that, for example, make up the art and technology tool set. [...]

Two or more autopoietic systems, say an artist and a technologist, each of which acts as a medium for the other, become mutually structurally coupled through the history of their reciprocal interactions. Events ('conducts', behaviours) in one system 'triggering perturbations' in another system bring the systems into an interlocking of interactions which is indistinguishable from what we call a 'consensual domain'. '[A] consensual domain is closed with respect to the interlocking conducts that constitute it, but is open with respect to the organisms or systems that realize it.'[6] It is this 'interlocked, mutually selecting, mutually triggering domain of state trajectories'[7] that is a collaboration, and I would consider that it is, in itself, an autopoiesis. [...]

We can think of collaboration in this way: the creation of a device or an artwork through a spiralling evolution brought about by the interactions of collaborators within an integrated system of feedforwards (being suggestions or enquiries), feedbacks (being responses) and adaptations. For those outside the collaboration, it would be seen as a self- generating autopoietic process. [...]

The individuals involved should be seen as discrete autopoieses that engage in feedback processes which bring a potential integration of the interests, intentions and skills of the individuals into what is a 'desiring machine' or system. The particular desiring machine is driven by the needs, desires, intentions and imaginations of the individual autopoietic entities within it. As a system it will, in the whole, be autopoietic, operating in a substratum of the consensual domain that we understand to be a society, and, in its construction as a jointly agreed project, becomes, in itself, a more narrowly determined consensual domain which is, thus, the collaboration. [...]

1 Jack Burnham, 'Systems Aesthetics', *Artforum* (September 1968).

2 Gilles Deleuze and Félix Guattari, 'Balance-Sheet Program for Desiring-Machines', trans. Robert Hurley, in *Semiotext(e)*, vol. 2, no. 3 (1977).

3 Ibid., 118–19, their emphasis.

4 Ibid., 121.

5 Ibid., 121, their emphasis.

6 Humberto R. Maturana, Biology of Language: The Epistemology of Reality', in George A. Miller and Elizabeth Lenneberg, eds, *Psychology and Biology of Language and Thought: Essays in Honour of Eric Lenneberg* (New York: Academic Press, 1978) 47.

7 Ibid., 39.

Stephen Jones, extracts from 'A Cultural Systems Approach to Collaboration in Art and Technology', *Proceedings of the 5th Conference on Creativity and Cognition* (London: Goldsmiths, University of London, 2005) 76–85.

Objects could be stretched in time, layered in time, scanned in time, filtered in time, metamorphosed and synchronized in time, in a matter of seconds

Sonia Landy Sheridan, 'Mind/Senses/Hand', 1990

GENERATIVE SYSTEMS

Richard Paul Lohse
Lines of Development//1943–84

To obtain a new operative basis it was necessary to systematize the media so that they could form logical sequences and would permit a multiplicity of operations. The result: variability and extensibility.

The colour series provides the law for formal expression, colour and form cancel each other out as opposites.

Anonymity of the media, non-limitation of the structural laws, relativity of the dimensions, extensibility and flexibility determine the expression.

The machine and the expression are developed at the same time, the method represents itself, it is the picture.

The picture field is a structural field.

The prerequisites for the development of flexible ordering systems are the identity of the pictorial media, of surface and surface boundaries, the anonymization and objectification of the structure, the congruence of the beginning and the end of the action.

The anonymous element is part and substance of a system of coordinates in which each element has an equal share of passivity and activity.

The individual expression lies in the choice of methods, in the control of preliminary conditions.

Simplicity is not produced by spontaneity but by the multiple superimposition, interpenetration and modification of the processes of development.

There is no definition of aesthetics without the definition of its social basis.

The task consists in developing systems that make transparent and combinable, flexible orders possible.

Technological reality is a fact that cannot be ignored. Identical with it is a vocabulary of media that is characteristic of this epoch, an instrumentarium of methods, systems, modes of behaviour, an arsenal of forms of expression that have already shaped the life of the epoch and will continue to shape it.

A social basis corresponds to every cultural expression, a cosmology to every aesthetic. In no other form of art do the media and methods of a global technological strategy find a legitimate expression as they do in constructive, logical, systematic or serial art, which is a sublimated and critical echo of the structures of civilization.

The forms of expression of a non-hierarchical society correspond to this society in the sphere of visual art: they are flexible, transparent, verifiable in method and in result.

It is as an instrument of cognition that art has a social value

Every technology has an appropriate sign alphabet that differs from its forerunner in structure, dimensions and motion. Forms of expression in art correspond to this global structure.

Flexibility is the counter-principle to monumentality.

No other epoch has experienced this onset of straight lines, direct connections, accumulations of identical elements resulting from the addition and division of the identical.

The possibility of repeating elements and facts mechanically is one of the identifying marks of this epoch.

Geometric artforms have a range extending from the esoteric to democratic orders.

There is no such thing as the language of geometry.

The serial principle is a radical democratic principle.

Richard Paul Lohse, extracts from 'Lines of Development' (1943–84); translated in *Richard Paul Lohse: Modulare und Serielle Ordnungen* (Zürich: Waser Verlag, 1984).

Iannis Xenakis
Free Stochastic Music//1965

Art, and above all, music has a fundamental function, which is to catalyse the sublimation that it can bring about through all means of expression. It must aim through fixations which are landmarks to draw towards a total exaltation in which the individual mingles, losing his consciousness in a truth immediate, rare, enormous and perfect. If a work of art succeeds in this undertaking even for a single moment, it attains its goal. This tremendous truth is not made of objects, emotions or sensations; it is beyond these, as Beethoven's Seventh Symphony is beyond music. This is why art can lead to realms that religion still occupies for some people.

But this transmutation of everyday artistic material which transforms trivial products into meta-art is a secret. The 'possessed' reach it without knowing its 'mechanisms'. The others struggle in the ideological and technical mainstream of their epoch which constitutes the perishable 'climate' and the stylistic fashion. Keeping our eyes fixed on this supreme meta-artistic goal, we shall attempt to define in a more modest manner the paths which can lead to it from our point of departure, which is the magma of contradictions in present music.

There exists a historical parallel between European music and the successive attempts to explain the world by reason. The music of antiquity, causal and deterministic, was already strongly influenced by the schools of Pythagoras and Plato. Plato insisted on the principle of causality, 'for it is impossible for anything to come into being without cause' (*Timaeus*). Strict causality lasted until the nineteenth century when it underwent a brutal and fertile transformation as a result of statistical theories in physics. Since antiquity the concepts of chance (*tyche*) , disorder (*ataxia*), and disorganization were considered as the opposite and negation of reason (*logos*), order (*taxis*), and organization (*systasis*). It is only recently that knowledge has been able to penetrate chance and has discovered how to separate its degrees – in other words to rationalize it progressively, without, however, succeeding in a definitive and total explanation of the problem of 'pure chance'.

After a time lag of several decades, atonal music broke up the tonal function and opened up a new path parallel to that of the physical sciences, but at the same time constricted by the virtually absolute determinism of serial music.

It is therefore not surprising that the presence or absence of the principle of causality, first in philosophy and then in the sciences, might influence musical composition. It caused it to follow paths that appeared to be divergent, but which, in fact, coalesced in probability theory and finally in polyvalent logic, which are kinds of generalization and enrichments of the principle of causality. The explanation of the world, and consequently of the sonic phenomena which surround us or which may be created, necessitated and profited from the enlargement of the principle of causality, the basis of which enlargement is formed by the law of large numbers. This law implies an asymptotic evolution towards a stable state, towards a kind of goal, of *stach as*, whence comes the adjective 'stochastic'.

But everything in pure determinism or in less pure indeterminism is subjected to the fundamental operational laws of logic, which were disentangled by mathematical thought under the title of general algebra. These laws operate on isolated states or on sets of elements with the aid of operations, the most primitive of which are the union, notated U, the intersection, notated n, and the negation. Equivalence, implication and quantifications are elementary relations from which all current science can be constructed.

Music, then, may be defined as an organization of these elementary operations and relations between sonic entities or between functions of sonic entities. We understand the first-rate position which is occupied by set theory, not only for the construction of new works, but also for analysis and better comprehension of the works of the past. In the same way a stochastic construction or an investigation of history with the help of stochastics cannot be carried through without the help of logic – the queen of the sciences, and I would even venture to suggest, of the arts – or its mathematical form algebra. For everything that is said here on the subject is also valid for all forms of art (painting, sculpture, architecture, films, etc.). [...]

[My article 'The Crisis of Serial Music' (1954)] served as a bridge to my introduction of mathematics in music. For if, thanks to complexity, the strict, deterministic causality which the neo-serialists postulated was lost, then it was necessary to replace it by a more general causality, by a probabilistic logic which would contain strict serial causality as a particular case. This is the function of stochastic science. 'Stochastics' studies and formulates the law of large numbers, which has already been mentioned, the laws of rare events, the different aleatory procedures, etc. As a result of the impasse in serial music, as well as other causes, I originated in 1954 a music constructed from the principle of indeterminism; two years later I named it 'Stochastic Music'. The laws of the calculus of probabilities entered composition through musical necessity.

But other paths also led to the same stochastic crossroads – first of all, natural events such as the collision of hail or rain with hard surfaces, or the song of cicadas in a summer field. These sonic events are made out of thousands of isolated sounds; this multitude of sounds, seen as a totality, is a new sonic event. This mass event is articulated and forms a plastic mould of time, which itself follows aleatory and stochastic laws. If one then wishes to form a large mass of point-notes, such as string *pizzicati*, one must know these mathematical laws, which, in any case, are no more than a tight and concise expression of a chain of logical reasoning. Everyone has observed the sonic phenomena of a political crowd of dozens or hundreds of thousands of people. The human river shouts a slogan in a uniform rhythm. Then another slogan springs from the head of the demonstration; it spreads towards the tail, replacing the first. A wave of transition thus passes from the head to the tail. The clamour fills the city, and the inhibiting force of voice and rhythm reaches a climax. It is an event of great power and beauty in its ferocity. Then the impact between the demonstrators and the enemy occurs. The perfect rhythm of the last slogan breaks up in a huge cluster of chaotic shouts, which also spreads to the tail. Imagine, in addition, the reports of dozens of machine guns and the whistle of bullets adding their punctuations to this total disorder. The crowd is then rapidly dispersed, and after sonic and visual hell follows a detonating calm, full of despair, dust and death. The statistical laws of these events, separated from their political or moral context, are the same as those of the cicadas or the rain. They are the laws of the passage from complete order to total disorder in a continuous or explosive manner. They are stochastic laws.

Here we touch on one of the great problems that have haunted human intelligence since antiquity: continuous or discontinuous transformation. The sophisms of movement (e.g. Achilles and the tortoise) or of definition (e.g. baldness), especially the latter, are solved by statistical definition; that is to say, by stochastics. One may produce continuity with either continuous or discontinuous elements. A multitude of short glissandi on strings can give the impression of continuity, and so can a multitude of *pizzicati*. Passages from a discontinuous state to a continuous state are controllable with the aid of probability theory. For some time now I have been conducting these fascinating experiments in instrumental works; but the mathematical character of this music has frightened musicians and has made the approach especially difficult.

Here is another direction that converges on indeterminism. The study of the variation of rhythm poses the problem of knowing what the limit of total asymmetry is, and of the consequent complete disruption of causality among durations. The sounds of a Geiger counter in the proximity of a radioactive source give an impressive idea of this. Stochastics provides the necessary laws.

Before ending this short inspection tour of events rich in the new logic, which were closed to the understanding until recently, I would like to include a short parenthesis. If glissandi are long and sufficiently interlaced, we obtain sonic spaces of continuous evolution. It is possible to produce ruled surfaces by drawing the glissandi as straight lines. I performed this experiment with *Metastasis* (this work had its premiere in 1955 at Donaueschingen). Several years later, when the architect Le Corbusier, whose collaborator I was, asked me to suggest a design for the architecture of the Philips Pavilion in Brussels, my inspiration was pinpointed by the experiment with *Metastasis*. Thus I believe that on this occasion music and architecture found an intimate connection. [...]

In line with [stochastic ideas], Michel Philippot introduced the calculus of probabilities into his painting several years ago, thus opening new directions for investigation in this artistic realm. In music he recently endeavoured to analyse the act of composition in the form of a flow chart for an imaginary machine. It is a fundamental analysis of voluntary choice, which leads to a chain of aleatory or deterministic events, and is based on the work *Composition pour double orchestre* (1960). The term imaginary machine means that the composer may rigorously define the entities and operating methods, just as on an electronic computer. In 1960 Philippot commented on his *Composition pour double orchestre*:

> If, in connection with this work, I happened to use the term 'experimental music', I should specify in what sense it was meant in this particular case. It has nothing to do with concrete or electronic music, but with a very banal score written on the usual ruled paper and requiring none but the most traditional orchestral

instruments. However, the experiment of which this composition was in some sense a by-product does exist (and one can think of many industries that survive only through the exploitation of their by-products).

The end sought was merely to effect, in the context of a work which I would have written independent of all experimental ambitions, an exploration of the processes followed by my own cerebral mechanism as it arranged the sonic elements. I therefore devised the following steps:

I. Make the most complete inventory possible of the set of my gestures, ideas, mannerisms, decisions and choices, etc., which were mine when I wrote the music.
2. Reduce this set to a succession of simple decisions, binary if possible; i.e. accept or refuse a particular note, duration or silence in a situation determined and defined by the context on one hand, and by the conditioning to which I had been subjected and my personal tastes on the other.
3. Establish, if possible, from this sequence of simple decisions, a scheme ordered according to the following two considerations (which were sometimes contradictory): the manner in which these decisions emerged from my imagination in the course of the work, and the manner in which they would have to emerge in order to be most useful.
4. Present this scheme in the form of a flow chart containing the logical chain of these decisions, the operation of which could easily be controlled.
5. Set in motion a mechanism of simulation respecting the rules of the game in the flow chart and note the result.
6. Compare this result with my musical intentions.
7. Check the differences between result and intentions, detect their causes, and correct the operating rules.
8. Refer these corrections back to the sequence of experimental phases, i.e. start again at l. until a satisfactory result has been obtained.

If we confine ourselves to the most general considerations, it would simply be a matter of proceeding to an analysis of the complexity, considered as an accumulation, in a certain order, of single events, and then of reconstructing this complexity, at the same time verifying the nature of the elements and their rules of combination. [...]

Thus appeared the phenomenon, a rather banal one, of autogeneration of complexity by juxtaposition and combination of a large number of single events and operations.

At the end of this experiment I possessed at most some insight into my own musical tastes, but to me, the obviously interesting aspect of it (as long as there is no error of omission!) was the analysis of the composer, his mental processes, and a certain liberation of the imagination.

The biggest difficulty encountered was that of a conscious and voluntary split in personality. On one hand was the composer who already had a clear idea and a precise audition of the work he wished to obtain, and on the other was the experimenter who had to maintain a lucidity which rapidly became burdensome in these conditions – a lucidity with respect to his own gestures and decisions. We must not ignore the fact that such experiments must be examined with the greatest prudence, for everyone knows that no observation of a phenomenon exists which does not disturb that phenomenon, and I fear that the resulting disturbance might be particularly strong when it concerns such an ill-defined domain and such a delicate activity.

Moreover, in this particular case, I fear that observation might provoke its own disturbance. If I accepted this risk, I did not underestimate its extent. At most, my ambition confined itself to the attempt to project on a marvellous unknown, that of aesthetic creation, the timid light of a dark lantern.

Iannis Xenakis, extracts from 'Free Stochastic Music' (1965), in Xenakis, *Formalized Music: Thought and Mathematics in Composition* (Bloomington: Indiana University Press, 1971); reprinted edition (Stuyvesant, New York: Pendragon Press, 1992) 1–11, 38–42.

Phivos-Angelos Kollias
Iannis Xenakis and Systems Thinking//2011

[...] Mathematical thinking in musical composition refers to the abstraction of sound elements, their quantification and the formalization of their relationships. That is to say, it is the rationalization of sound's control. Although the use of mathematics as a compositional tool does not necessarily suggest aesthetic values, or particular modes of perception, the application of stochastics [procedures based on random variables and probability], via a cybernetic epistemology, opens up a new field of musical creation and experience. [...]

As Francisco Valera suggests, one of the basic aims of cybernetics was the attempt to organize a science of the mind. [*Invitation aux sciences cognitives*, 1996] From this perspective, what had previously been monopolized by philosophers and psychologists became a subject of study for interdisciplinary teams, who would search for the underlying processes of the mind and describe them in terms of explicit mechanisms and mathematical formalizations.

This may help explain the reason for Xenakis' interest in systems thinking. As described by Makis Solomos, in Xenakian aesthetics, which has a clearly anti-romantic attitude, the focus is no longer on the 'heart' but the 'brain'; no longer on sentiment but the sensory. ['Les "opérations mentales de la composition"', *Intellecta* (2008)] For Xenakis logic, rather than beauty, now rules the arts.

Another reason is that for Xenakis art was not a matter of cultivated minds – an attitude that led away from humanity's biological foundations and would result, as he put it, in a 'sterile desert'. Instead, he extends the sensorial aspect of music so as to form a direct connection between human biological nature and intelligence.

A third reason is Xenakis' tendency towards universality. He envisages a new kind of musician, who possesses 'a sort of universality, but one based upon, guided by and oriented toward forms and architectures',[1] and who is conversant with mathematics, logic, physics, chemistry, biology, genetics, palaeontology, the human sciences, history. [...]

Another important aspect of systemic thought, also connected with universality, is the concept of isomorphism. According to Ludwig von Bertalanffy, 'the consequence of the existence of general system properties is the appearance of structural similarities or isomorphisms in different fields. There are correspondences in the principles that govern the behaviour of entities that are intrinsically widely different.' [*General Systems Theory*, 1968]

Similarly, Xenakis states that 'any theory or solution given to one level can be assigned to the solution of problems on another level. ... [Q]uestions having to do mainly with orchestral sounds ... find a rich and immediate application as soon as they are transferred to the microsound level ... All music is thus automatically homogenized and unified.' [...]

At the base of [the psychologist and cybernetician] Ross Ashby's interpretation of cybernetics is the concept of difference:

1. The difference between two discernible things.

2. The difference between one thing and its change to another. [*An Introduction to Cybernetics*, 1956] [...]

Xenakis describes what he calls Markovian stochastic music, starting with a basic hypothesis: 'All sound is an integration of grains, of elementary sonic particles, of sonic quanta. Each of these elementary grains has a threefold nature: duration, frequency and intensity.' He explains his hypothesis with a metaphor: 'A complex sound can be imagined as a multicoloured firework, in which each point of light appears and instantaneously disappears against a black sky.' In order to model any complex sound (viewing Xenakis' hypothesis from a cybernetic perspective), the only three parameters that interest us are duration, frequency and intensity. [...]

In order to describe sound in finite steps, Xenakis takes chunks of time of unchanging length, so as to simplify the model and keep only two changing parameters. In this way, in every instance, the state of a grain [of sound] can be described by a vector (f, g) where the mapping (the correspondence) between the frequencies and the intensities can be single-valued or many-valued. That is to say, in the former case a frequency may correspond to only one intensity, while in the latter case a frequency may correspond to many possible intensities. [...]

As Ashby does in his cybernetics, Xenakis first describes determinate transformations and later introduces stochastic ones. A determinate transformation is closed: all the elements of the transformation are predefined; and it is single-valued: each operant is converted to one, and only one, transform. [...] Xenakis also suggests another use of cybernetic representation, where the elements of the transformation may represent, instead of screens, other musical elements: notes, rhythmic values, textural qualities, timbres or 'concrete music characters'. [...]

Interpreting Xenakis' music through systems thinking, we can more profoundly understand his methodologies and resulting aesthetic values. This can inform us more about his music than a simple approach through mathematical formulae. Furthermore, a knowledge of the evolution of these theories can contribute to new applications in music.

1 [This and subsequent quotations are taken from Iannis Xenakis, *Formalized Music: Thought and Mathematics in Composition* (New York: Pendragon Press, 1992).]

Phivos-Angelos Kollias, extracts from 'Iannis Xenakis and Systems Thinking', Proceedings of the Iannis Xenakis International Symposium (London: Southbank Centre, 2011). (www.gold.ac.uk/ccmc/xenakis-international-symposium)

Manfred Mohr
Statement//1971

Accepting that creative work is an algorithm which represents a human behaviour in a given situation, it is natural to ask: how is such an algorithm built up, and which precise mathematical laws could be extracted for later use in different circumstances? If one is now curious enough to look for his own aesthetic parameters, he is ready to engage in an interesting line of research. These considerations led me to use the computer as a PARTNER in my work.

The first step was an extended analysis of my own paintings and drawings from the last ten years. It resulted in a surprisingly large amount of regularities, determined of course by my particular aesthetic sense, through which I was able to establish a number of basic elements that amounted to a rudimentary syntax. After representing these basic constructions through a mathematical formalism, and setting them up in an abstract combinatorial framework, I was in a position to realize all possible representations of my algorithms.

Since the most important point in applying a computer to solve aesthetic problems is the MATERIALGERECHTE[1] use of this instrument, the research therefore should assume that old techniques of drawing and imagination are not to be imposed on the machine (although this would be possible), but should develop a priori a vocabulary which integrates the computer into the aesthetic system.

Computer graphics in general are conditioned by four basic premises:

1. A PRECISE idea of an aesthetic problem.

2. The need to break this idea into parts which could be reassembled as a program.

3. A steady control of the computing process to take full advantage of the MACHINE – HUMAN dialogue.

4. The need for the logic of the events to become perceptible.

The logic built into a program makes it possible to create a nearly infinite number of new situations. This is very important since the creation of a form is limited a priori by its author's characteristics, of which he may be conscious or unconscious. It means that the exploration of a new idea leads sooner or later to a repetition which can be avoided by resorting to a computer once the basic parameters have been formulated. As it is possible to conceive the logic of a construction but not all its consequences, it is nearly an imperative to rely on a computer to show this large variety of possibilities; a procedure which may lead to different and perhaps more interesting answers, lying of course outside of normal behaviour but not outside of the imposed logic.

At this point a new problem appears: how to choose what is to be kept and what is to be rejected?

My aesthetic criteria were determined by a decision not to create single forms but sets of forms. The basic parameters are the relationships between the forms, and no aesthetic value is associated to particular forms. Within this context it is possible to ignore the former 'good' and 'bad', and aesthetic decisions can be based on WERTFREIE[2] procedures, where the totality represents a 'quality of a quantity'. The fundamental consequence of this attitude is, that after a period of tests, modifications of the logic and parameter exchanges, all possible results of a program have to be rigorously accepted as final answers.

Computer graphics is a young and new way of aesthetic communication; it integrates human thinking, mechanical handling, logic, new possibilities of drawing and incorruptible precision of drawing – a new DUKTUS[3]

The concentration which is necessary to establish a logic (writing a program – that means to give a definition of all instructions that have to be done in the machine) will reflect itself in the result as a clear construction which could be understood by everybody, and there will be less and less mystical barriers behind which the artist can hide himself.

1 *Materialgerecht* – Working or using a material only in the way which is basic to the material.

2 *Wertfrei* – Decisions, where the knowledge is neither based nor conditioned by any values.

3 *Duktus* (Latin) – German for 'handwriting'. Individual peculiarity of the drawing material.

Manfred Mohr, Statement (1971), in *Manfred Mohr. Computer Graphics – Une Esthétique Programmée* (Paris: ARC/Musée d'art moderne de la Ville de Paris, 1971) 36–40.

Sonia Landy Sheridan
Mind/Senses/Hand: The Generative Systems Program at the Art Institute of Chicago, 1970–80//1990

The 1960s were tumultuous years in Chicago. There was enough charged negative and positive energy in the air to move the most inert of us to creative activity. It was in this climate that a new program, Generative Systems, was born at the School of the Art Institute of Chicago. [...] [A]t first we taught extensions of traditional art processes, but later we developed a full program of investigation into the transformative process occurring in art as a result of the impact of the communications revolution on the society at large. Generative Systems was a research centre; a resource and energy bank; a self-generating centre where communication tools came and went while people remained; a nurturing ground for the Electronic Printout Systems (ESP); an extension into the future of photography, drawing, textiles, and so on; a time machine from instant real-time back to mechanical time; an attitude; an interactive force between industry, education and the public; and finally a viable alternative to the present art education system.

Although Generative Systems courses were formally begun in the 1970s, they were rooted in my work at the Institute in the 1960s when I taught [...] the basic art school foundation courses. They were influenced for the most part by Bauhaus teachings. [...] [H]owever, my main educational resource was the highly pictorial journal *Scientific American*. It was not until I worked with high-speed communications tools that Moholy-Nagy's *Vision in Motion* (1947) took on a real significance for me. The new communication imaging systems validated his perceptions. [...]

For a decade, from 1967 to 1977, we were occupied with exploring many communications systems [...]. The communications technology that emerged in the 1960s validated the dreams of the most imaginative minds. Objects could be stretched in time, layered in time, scanned in time, filtered in time, metamorphosed and synchronized in time, in a matter of seconds, on the new electronic copiers, telecopiers and computers, with their moving lights, lenses, thermal and/or steel rollers and electronic gates.

During 1969 and 1970 I created a body of work with copiers and their by-products that led to my becoming artist-in-residence at 3M's Color Research Laboratory with Douglas Dybvig, laboratory director and inventor of 3M's Color-in-Color photocopying machine. This gave me the foundation needed to establish the first Generative Systems course in 1970 [...]. Then in the two ensuing years Generative Systems became an energy bank, tele-link-up, exchange centre, city nerve centre, public relations centre and interdepartmental link-up providing events, activities and performances. By the seventh year, the courses called Process I and Process II were created. This was partly as a result of my renewed exposure to scientists in 1976 as an artist-in-residence at 3M's Central Research Laboratories [...].

Process I [...] examin[ed] energy for imaging manually, mechanically, electronically and photonically. Process II gave the student an opportunity to pull apart and examine dozens of communication machines, such as high-speed copiers, video recorders and computers. One of the teaching assistants was Greg Gundlach, who in the process began research for a three-dimensional photographic system that he would after graduation patent as Z-Tranz.

Computer Graphics was finally made into a course in the late 1970s, when I obtained a 4K Radio Shack computer with a thermal silver paper printer. In a few months a Z-80 computer was assembled from a kit by a graduate student teaching assistant, John Dunn, who in the process was setting the basis for the first computer graphic system for artists, his SLIDEMASTER, which became EASEL and then Lumena (Time Arts Inc.). [...]

[S]ince the main emphasis of both courses was not on making 'Art', another course seemed necessary. It was not until the end of the 1970s, however, that we were ready for a course that would permit all aspects of the artistic process to function as a unity. [...]

The course called Homography was created for just this purpose – to decide what to do with our new-found knowledge. [We used] tools from a whole spectrum of eras: the pen, pencil and brush; the camera; the copier and video recorder and computer [...] to create problems that did not yet exist [...], to explore the conceptual, artistic and scientific implications of the area. By the end of the semester I had created nine new lessons, nine ways of visualizing time through the use of manual, mechanical, electronic and photonic tools. George Kubler's *The Shape of Time* (1962) and Moholy-Nagy's *Vision in Motion* were being realized not merely in film, video and sound, which were by nature time studies, but also in what are normally considered 'still arts': drawing, painting, printmaking and photography.

Moving-time and stopped-time imaging systems are interchangeable, but it was not until the availability of electronic photo/print processes in the 1960s that images, not merely of our imagination, could be stopped in time by simple accessible systems. Photographers knew of Harold Edgerton's pioneering work in stopped-time images, but fields outside photography, film, video and sound did not deal with multiple dimensions. [...]

Process I and II provided the minimal experience with technology needed to pursue the development of Generative Systems. Homography was the course designed to permit two aspects of the creative process – personal/inner and objective/outer – to function as a unity. It was a search for the poetry of the process. It was an attempt to find the aesthetics and meaning underlying the shift from tools of one kind of time to tools of another kind of time. This was a complex process, and in my own case, since I was learning along with the students, it could be achieved only through the total integration of my own work and production with that of the classroom needs.

I have had many challenging discussions with splendid artist/educators who found my system to be dangerous, to say the least. My choice for this integrative process can best be explained by recognizing that my context, in Chicago at the time, seemed to demand a democratic, decentralized program with the support of people in industry, artists, those in educational institutions and a host of other people. My personal philosophy and my desire to integrate a first-generation creative process into art production and training, in synchronization with social and technological development, led me to no other conclusion. I could find no other acceptable alternative for the Chicago art school context. The time and place seemed to demand a fluid, non-dictatorial context. Yet the program had to be based on solid, objective discovery rooted in a knowledge of art history and scientific/artistic discovery. [...]

In retrospect, some aspects of the program that are applicable to other quite different educational approaches appear to be the following:

1. In studying nature's processes, we discover basic common underlying unities, structures and patterns. Not only is the river delta a 'new landscape', it is

a dendritic pattern in our hands, in a heat-pressured copier sheet, in the positive charge of Xerox electrostatic toner on a selenium plate, or in the pressure of a foot in the mud. If one moves one's foot in a flow pattern one creates a wave; if one runs a hot iron over dye-coated copier paper one creates again a wave form.

2. Artists can work with scientists to study nature's processes as well as its products. The underlying structures of systems can be explored by both artist and scientist separately or in tandem.

3. Artists can use mind, senses and hand to examine and explore energy sources to enhance their awareness. Light, heat, magnetics, electrostatics and sound transmission can be playfully and systematically explored, even by using children's science books, manually, mechanically, electronically, photonically and biologically.

4. Our perceptions of time and space are altered as we change our tools from manual to electrostatic to photonic and biologic or any combination of these. We can ride in time through past, present and future via our choice of tools.

5. Stopped-time and moving-time systems, when playfully and systematically explored, can reveal new ways of perceiving ourselves in time and space. We ride in space/time through our choice of system.

6. Ways of visualizing time can be studied subjectively/objectively, inwardly/outwardly, as was colour by Johannes Itten [in *The Art of Colour*, 1961]. [...]

7. Artists can create their own sophisticated tools and thus affect society; or they can separately or simultaneously pick up tools discarded by the society for creative experimentation.

8. No tools are too outmoded for creative use. No tools are too new for creative use. An artist can create with any tools, but certain tools are linked to the dynamics of social and technological development and open up vistas in a special way.

9. Complex systems can be invented even by young students. One key is to direct the student to look at the mirror image of a problem such as three-dimensional photo-imaging with the grid on the camera rather than on the receptor.

10. There need not be artist/technicians who only create tools or operate them while others, artists proper, create art with these tools. Timing, need and human preference and adaptability create a wide range of choices in a single lifetime. Specialization is for researchers and robots.

11. Networks of individuals can build an energy bank, and with it creative excitement, even in static academic environments.

12. An artist/educator who openly and freely provides a visible model of creating through teaching, exploration and research can produce a body of artists capable of creating their own tools, or creating their own art centres, or inventing the next stage of multidimensional visualization. The product does not disappear

in the process. Process and product are two components of one system – the creative process.

13. Artists can explore and record the processes of evolving, moving systems as well as of static ones.

14. Generative Systems does not have to be only the closed system of its historical past. It can be an open/closed system or a mind/body system. [...]

The Generative Systems program at the Institute was not a closed system or a variation on a theme. It was an open system, an ever-changing system, in which the machines would come and go, but the humans would remain the constant factor. Courses would not be named for a specific and therefore static technological process [...] but rather for a dynamic process encompassing change, metamorphosis, inconsistency and chaos. In the process, the mind/body of the human being could create closed systems and open systems, neither one negating the other, but, rather, each complementing the other in a process of continual becoming. [...]

The Generative Systems program was just one way, in one place and at one time, to tackle common problems of creativity in art, science and technology. Perhaps Generative Systems' ten-year existence in an institution was validated by its graduates, who invented new systems for society, set up new learning centres, created new artforms and influenced yet another generation of artists. [...]

Sonia Landy Sheridan, extracts from 'Mind/Senses/Hand: The Generative Systems Program at the Art Institute of Chicago, 1970–1980', *Leonardo*, vol. 23, no. 2–3 (1990) 175–81 [footnotes not included].

Brian Eno
Generating and Organizing Variety in the Arts//1976

A musical score is a statement about organization; it is a set of devices for organizing behaviour toward producing sounds. That this observation was not so evident in classical composition indicates that organization was not then an important focus of compositional attention. Instead, the organizational unit (be it the orchestra or the string quartet or the relationship of a man to a piano) remained fairly static for two centuries, while compositional attention was directed at using these given units to generate specific results by supplying them with specific instructions. [...]

I shall be using the term *variety* frequently in this essay and I should like to attempt some definition of it now. It is a term taken from cybernetics (the science of organization) and it was originated by W.R. Ashby.[1] The *variety* of a system is the total range of its outputs, its total range of behaviour. All organic systems are probabilistic: they exhibit variety, and an organism's flexibility (its adaptability) is a function of the amount of variety that it can generate. Evolutionary adaptation is a result of the interaction of this probabilistic process with the demands of the environment. By producing a *range* of outputs evolution copes with a *range* of possible futures. The environment in this case is a *variety-reducer* because it 'selects' certain strains by allowing them to survive and reproduce, and filters out others. But, just as it is evident that an organism will (by its material nature) and must (for its survival) generate variety, it is also true that this variety must not be unlimited. That is to say, we require for successful evolution the transmission of *identity* as well as the transmission of *mutation*. Or conversely, in a transmission of evolutionary information, what is important is not only that you get it right but also that you get it slightly wrong, and that the deviations or mutations that are useful can be encouraged and reinforced.

My contention is that a primary focus of experimental music has been toward its own organization, and toward its own capacity to produce and control variety, and to assimilate 'natural variety' – the 'interference value' of the environment. Experimental music, unlike classical (or avant-garde) music, does not typically offer instructions toward highly specific results, and hence does not normally specify wholly repeatable configurations of sound. It is this lack of interest in the *precise* nature of the piece that has led to the (I think) misleading description of this kind of music as *indeterminate*. I hope to show that an experimental composition aims to set in motion a system or organism that will generate unique (that is, not necessarily repeatable) outputs, but that, at the same time, seeks to limit the range of these outputs. This is a tendency toward a 'class of goals' rather than a particular goal, and it is distinct from the 'goalless behaviour' (indeterminacy) idea that gained currency in the 1960s.

I should like to deal at length with a particular piece of experimental music that exemplifies this shift in orientation. The piece is Paragraph 7 of *The Great Learning* by Cornelius Cardew,[2] and I have chosen this not only because it is a compendium of organizational techniques but also because it is available on record. [...] Implicit in the score is the idea that it may be performed by *any* group of people (whether or not trained to sing). The version available on record is performed by a mixed group of musicians and art students, and my experience of the piece is based on four performances of it in which I have taken part.

Cardew's score is very simple. It is written for any group of performers (it does not require trained singers). There is a piece of text (from Confucius) which is divided into 24 separate short phrases, each of one to three words in length.

Beside each phrase is a number, which specifies the number of repetitions for that line, and then another number telling you how many times that line should be sung loudly. The singing is mostly soft.

All singers use exactly the same set of instructions. They are asked to sing each line of the text the given number of times, each time for the length of a breath, and on one note. The singers start together at a signal, and each singer chooses a note for the first line randomly, staying on it until the completion of the repetitions of the line.

The singer then moves on to the next line, choosing a new note. The choice of this note is the important thing. The score says: 'Choose a note that you can hear being sung by a colleague. If there is no note, or only the note you have just been singing, or only notes that you are unable to sing, choose your note for the next line freely. Do not sing the same note on two consecutive lines. Each singer progresses through the text at his own speed.'

A cursory examination of the score will probably create the impression that the piece would differ radically from one performance to another, because the score appears to supply very few *precise* (that is, quantifiable) constraints on the nature of each performer's behaviour, and because the performers themselves (being of variable ability) are not 'reliable' in the sense that a group of trained musicians might be. The fact that this does not happen is of considerable interest, because it suggests that *somehow a set of controls that are not stipulated in the score arise in performance* and that these 'automatic' controls are the real determinants of the nature of the piece. [...]

In summary, then, the generation, distribution and control of notes within this piece are governed by the following: one specific instruction ('do not sing the same note on two consecutive lines'), one general instruction ('sing any note that you can hear'), two physiological factors (tone-deafness and transposition), two physical factors (beat frequencies and resonant frequency), and the cultural factor of 'preference'. Of course, there are other parameters of the piece (particularly amplitude) that are similarly controlled and submit to the same techniques of analysis, and the 'breathing' aspects of the piece might well give rise to its most important characteristic – its meditative calm and tranquillity. But what I have mentioned above should be sufficient to indicate that something quite different from classical compositional technique is taking place: the composer, instead of ignoring or subduing the variety generated in performance, has constructed the piece so that this variety is really the substance of the music.

Perhaps the most concise description of this kind of composition, which characterizes much experimental music, is offered in a statement made by the cybernetician Stafford Beer. He writes: 'Instead of trying to specify it in full detail, you specify it only somewhat. You then ride on the dynamics of the system in the direction you want to go.'[3] In the case of the Cardew piece, the 'dynamics of the system' is its interaction with the environmental, physiological and cultural climate surrounding its performance. The English composer Michael Parsons provides another view on this kind of composition:

> The idea of one and the same activity being done simultaneously by a number of people, so that everyone does it slightly differently, 'unity' becoming 'multiplicity', gives one a very economical form of notation – it is only necessary to specify one procedure and the variety comes from the way everyone does it differently. This is an example of making use of 'hidden resources' in the sense of natural individual differences (rather than talents or abilities), which is completely neglected in classical concert music, though not in folk music.[4]

This movement toward using natural variety as a compositional device is exemplified in a piece by Michael Nyman called *1-100* (Obscure 6). In this piece, four pianists each play the same sequence of one hundred chords descending slowly down the keyboard. A player is instructed to move on to his next chord only when he can no longer hear his last. As this judgement is dependent on a number of variables (how loud the chord was played, how good the hearing of the player is, what the piano is like, the point at which you decide that the chord is no longer audible), the four players rapidly fall out of sync with one another. What happens after this is that unique and delicate clusters of up to four different chords are formed, or rapid sequences of chords are followed by long silences. This is an elegant use of the compositional technique that Parsons has specified, not least because it, like the Cardew piece, is extremely beautiful to listen to – a factor that seems to carry little critical weight at present. [...]

Given [my] reservation about polarizing musical ideas into opposing camps, I should now like to describe two organizational structures. My point is not that classical music is one and contemporary music the other, but that each is a group of hybrids tending toward one of the two structures. At one extreme, then, is this type of organization: a rigidly ranked, skill-oriented structure moving sequentially through an environment assumed to be passive (static) toward a resolution already defined and specified. This type of organization regards the environment (and its variety) as a set of emergencies and seeks to neutralize or disregard this variety. An observer is encouraged (both by his knowledge of the ranking system and by the differing degrees of freedom accorded to the various parts of the organization) to direct his attention at the upper echelons of the ranks. He is given an impression of a hierarchy of value. The organization has the feel of a well-functioning machine: it operates accurately and predictably for one class of tasks but it is not adaptive. It is not self-stabilizing and does not easily assimilate change or novel environmental conditions. Furthermore, it requires a particular type of instruction in order to operate. In cybernetics this kind of instruction is known as an *algorithm*. Stafford Beer's definition of the term is 'a comprehensive set of instructions for reaching a known goal'; so the prescription 'turn left at the lights and walk twenty yards' is an algorithm, as is the prescription 'play a C-sharp for a quaver followed by an E for a semiquaver.' It must be evident that such specific strategies can be devised only when a precise concept of form (or identity, or goal, or direction) already exists, and when it is taken for granted that this concept is static and singular.

Proposing an organizational structure opposite to the one described above is valueless because we would probably not accord it the name *organization*: whatever the term does connote, it must include some idea of constraint and some idea of identity. So what I shall now describe is the type of organization that typifies certain organic systems and whose most important characteristics hinge on this fact: that changing environments require adaptive organisms. Now, the relationship between an organism and its environment is a sophisticated and complex one, and this is not the place to deal with it. Suffice it to say, however, that an adaptive organism is one that contains built-in mechanisms for monitoring (and adjusting) its own behaviour in relation to the alterations in its surroundings. This type of organism must be capable of operating from a different type of instruction, as the real coordinates of the surroundings are either too complex to specify, or are changing so unpredictably that no particular strategy (or specific plan for a particular future) is useful. The kind of instruction that is necessary here is known as an *heuristic*, and is defined as 'a set of instructions for searching out an unknown goal by exploration, which continuously or repeatedly evaluates progress according to some known criterion.'[6] To use Beer's example: if you wish to tell someone how to reach the top of a mountain that is shrouded in mist, the heuristic 'keep going up' will get him there. An organism operating in this way must have something more than a centralized control structure. It must have a responsive network of subsystems capable of autonomous behaviour, and it must regard the irregularities of the environment as a set of opportunities around which it will shape and adjust its own identity. [...]

1 [footnote 2 in source] W. Ross Ashby, *An Introduction to Cybernetics* (1956) reprinted edition (London: University Paperbacks, 1964).

2 [3] Each paragraph corresponds to one in the Confucian classic of the same title.

3 [6] Stafford Beer, *Brain of the Firm: The Managerial Cybernetics of Organization* (London: Allen Lane, 1972) 69.

4 [7] Michael Parsons, quoted in Michael Nyman, *Experimental Music* (New York: Schirmer, 1974).

5 [9] Stafford Beer, *Brain of the Firm*, op. cit., 305.

6 [10] Ibid., 306.

Brian Eno, extracts from 'Generating and Organizing Variety in the Arts', *Studio International*, no. 193 (November/December 1976); reprinted in *Audio Culture: Readings in Modern Music*, ed. Christoph Cox and Daniel Warner (London: Bloomsbury Academic, 2004) 226–33.

Michael Joaquin Grey
Statement//*c*. 2004

With the development of super computers by the late 1980s it was possible to model a system close to the order of complexity of natural systems, a new territory for the art of observation. I started to record the ontogeny (development) of information: experience, observation, description, explanation and exploitation of form in this new iterative space. Just as Leeuwenhoek looked at cells (biological) for the first time, or Kepler looked at the macrocosmos, I saw the rare opportunity to experience first hand the hubris and problems of the early development of discovery. I worked with Randy Huff to develop proprietary software to visualize some of the first neural networks and genetic algorithms capable of autonomous learning and behaviour. I was interested in recapitulating the dreams of causality that were part of exploring any new frontier.

I found the language to describe and explain the behaviour of information and Artificial Life programs very challenging linguistically. I eventually developed the Citroid System and ZOOB modelling system to have a manipulative [design set] to share and express the unity of complexity and dynamics of information, micro, macro and biological behaviour. I found the linguistic syntax limited to modelling spatial syntax and complexity. Prior to the Citroid System and ZOOB, there were only two variations of manipulative modelling; stereotonic modelling, or stacking, based on the development of the city, the brick, and tectonic, based on engineering from the industrial revolution to Buckminster Fuller. My modelling system is dynamic, based on how the body works, micro, macro and information behaviour. This was the basis for the Citroid System and ZOOB, with body empathy and self-similarity, from molecular behaviour (DNA and protein formation) to the scale of the joints and anatomy of the human body (animation), to celestial formations (network and macro models).

Michael Joaquin Grey, Information statement (*c*. 2004) (http://www.citroid.com)

Christa Sommerer and Laurent Mignonneau
Art as a Living System//1999

During the six years of our collaboration, we have worked at the borderline of art and biology and have used biological principles to create interactive artworks. [...]

In 1992 we developed the concept of natural interfaces and evolutionary image processes linked to interaction. We started working with evolutionary biology and became increasingly intrigued by how natural evolution and the processes of nature can function as tools of creation. [...]

Based on the insight that interaction per se and the interrelation between entities are the driving forces behind the structures of life, we investigated interaction and the creative process. Creation is no longer solely understood as an expression of the artist's inner creativity, but instead becomes an intrinsically dynamic process. Linking the interaction of human observers (visitors) directly to the dynamic and evolutionary image processes of an artwork allows us to create artworks that are under constant change and development.

We believe that interaction should not be linear but instead feel like a journey. The more one engages in interaction, the more one learns about it and the more one can explore it. We call this principle non-linear or multi-layered interaction: interaction should be easy to understand at the very beginning but also rich so that the visitor is able continuously to discover different levels of interactive experiences. [...]

One of our first interactive computer installations to use a natural interface, instead of then common devices such as joysticks, mouse, trackers or other technical interfaces, was *Interactive Plant Growing* (1993). In this piece, living plants function as the interface between the human visitor and the artwork. [...]

[W]e became increasingly inquisitive about the process of creation itself. Artificial Life (A-Life), a research field invented by scientist Christopher Langton at the Santa Fe Institute, proved capable of producing processes of nature within a machine (computer environment) and allowed computer programs to evolve over time. This enabled the development of processes and patterns that are no longer predictable or 'handmade'.

Fascinated by the idea of creation through evolution, not as a scientific simulation or mimicry of nature but as an investigation into the creative process itself, we studied the possibilities of applying A-Life principles to art projects.

Natural evolution has brought about a vast variety of forms and structures in nature, so it seemed reasonable to us that artificial evolution could function as a mechanism of the visual creation process. Also inspired by John Cage's use of chance procedures in his musical compositions, we began to introduce a combination of interaction and artificial evolution to our works.

In 1994 we started to collaborate with Tom Ray, A-Life scientist and creator of the 'Tierra' system. During this collaboration we developed the interactive computer installation *A-Volve*, which allowed visitors to create A-Life (in the form of artificial creatures) and to interact with it. Artificial creatures are basically computer-generated forms that display life-like behaviour and interact with each other as well as with their environment. *A-Volve* features A-Life principles in the birth, creation, reproduction and evolution of its artificial creatures.

In an interactive real-time environment, *A-Volve* visitors interact with virtual creatures in a water-filled glass pool. These virtual creatures are products of evolutionary rules and are influenced by human creation and interaction. Designing any kind of shape and profile with his/her finger on a touch-screen, the visitor creates virtual 3-D creatures that are 'alive' and swim in the real water of the pool. The movement and behaviour of the virtual creature are decided by its form – that is, how the viewer designs it via the touch-screen. [...]

Cross-over between the genetic strings of the parent creatures, as well as the mutation and selection of fitter creatures, provide a simulation of reproduction mechanisms found in nature. Newborn offspring also live in the pool, interacting with visitors and other creatures.

Laurent designed algorithms to ensure smooth and natural movements and 'animal-like' behaviour of the creatures. None of the creatures are pre-designed; they are all born exclusively in real time through the interaction of the visitors and the mating processes of the creatures themselves. [...]

Since the genetic code of the offspring is carried from generation to generation and the system emphasizes selection of fitter creatures, the code is able to evolve over time toward fitter creatures. Although evolution can take place by itself without outside influence, the system is designed in such a way that the visitor and his/her interaction and creation of forms will significantly influence the evolutionary process. The visitors act as a kind of external selection mechanism. [...]

All of *A-Volve*'s entities – the images, the forms and the graphical environment – change continuously, as does the audience itself, their imaginations, the ways they conceive and draw forms and how they interact with them. Human-creature interaction itself becomes a creative process. The social interaction between the viewers and the virtual world is essential to the creation of the work itself. We think of *A-Volve* as a complex system in which, as in quantum physics, the entities transform their states according to probability patterns. This system is like an interconnected, intrinsically dynamic web of movement, interaction and transformation of particles and entities. [...]

In 1996 we began to study the building blocks of visual creation and investigated how simple structures can result in complex-looking shapes and forms through genetic manipulations. We developed *GENMA* (Genetic Manipulator) – an interactive installation that allows visitors to create, manipulate and explore the genetic design of artificial creatures – for the Ars Electronica Centre (AEC) in Linz, Austria, as part of a permanent exhibition.

GENMA is a kind of dream machine enabling us to manipulate artificial nature on a micro scale: abstract, amoeboid, artificial 3-D forms and shapes. Principles of A-Life and genetic programming are implemented in this project, which allows visitors to manipulate the virtual genes of the creatures in real time. [...]

On a visual level, *GENMA* further explores the concept of 'natural' or 'open-ended' design – design that is not pre-fixed or controlled by the artists but that represents the degree of interest and interaction of each visitor. Each visitor creates the forms he/she wants to see, aided by artificial genetics, mutation and manipulation. One could even say that the audience become artists themselves, using the power and possibilities of the installation's tools.

In 1997 we extended the concept of *GENMA* a step further and implemented the principles of open-ended design in an installation called *Life Spacies* [...], an interaction and communication environment in which remotely located visitors via the Internet and onsite visitors at the installation in Tokyo can interact with each other through evolutionary forms and images.

Through the *Life Spacies* web page, people all over the world interact with the system as well. By simply typing and sending an e-mail message to the *Life Spacies* website one can create one's own artificial creature.

We developed a special text-to-form coding system that allows us to translate text into genetic code. In a way similar to what occurs in nature, letters, syntax and sequencing of the text are used to code certain parameters in a creature's design. Form, shape, colour, texture and the number of limbs are influenced by text parameters. As there is great variation in the texts sent by different people, the creatures themselves also vary greatly in their appearance, thus resulting in unique creatures for each participant.

As soon as the message is sent, the produced creature starts to live in the *Life Spacies* environment at the ICC museum, where on-site visitors can interact with it directly through touch. [...]

The artificial species can be created in one of two different ways:

Through incoming e-mail messages. A text-to-form editor creates the genetic code for each creature: one message equals one creature; complex text messages create complex creatures; and different levels of complexity within the text represent different species.

Through reproduction of the creatures themselves. Reproduction helps the creatures propagate their genotype in the system so they can form groups of different species.

Life Spacies is also based on the idea of evolutionary design – the result is not predetermined by the artist but depends solely on the interaction of the visitors and the evolutionary process. Only the messages e-mailed from people throughout the world and the reproduction and evolution of the creatures themselves determine how the creatures look and how they behave. One can therefore never really predict how the work will evolve and what kind of creatures will emerge. Its evolution depends on how many people send messages, how complex these messages are and how the creatures reproduce among themselves and through the selection of visitors at the museum.

Life Spacies is a system where interaction, interrelation and exchange happens on human-human, human-creature, creature-creature, and human-environment, creature-environment and real life–A-Life levels. [...]

The interaction rules are non-deterministic and multilayered; our aim was to create an open-ended system in which each entity – whether real life or A-Life, whether actually present (visitors at the ICC Museum) or virtually present (the users on the Internet or the creatures as code) – are equally important components of a complex, life-like system. [...]

Interactivity and A-Life teach us to rethink our definition of art, broadening our view by allowing us to integrate personality, variety, processes of nature and new perspectives on art and life. As the images in our installations are not static, pre-fixed or predictable, they become living processes themselves, representing the influences of the viewers' interactions and the internal principles of variation, mutation and evolution. The image processes are no longer reproducible but continuously changing and evolving. Such artwork can therefore be considered a living system itself, representing the relationship and interaction between life and A-Life.

Christa Sommerer and Laurent Mignonneau extracts from 'Art as a Living System', *Leonardo*, vol. 32, no. 3 (June 1999) 165–73 [footnotes not included].

Ken Rinaldo
Autopoiesis//2000

My interdisciplinary media art installations look to the intersection between natural and technological systems. Integration of the organic and electro-mechanical elements asserts a confluence and co-evolution between living and evolving technological material. I am fascinated with and encouraged by humankind's struggle to evolve technological systems that move toward intelligence and autonomy, which are modelled from our current conceptions of the natural. My artworks are influenced by theories on living systems, artificial life, interspecies communication and the underlying beauty and pattern inherent in the nature and organization of matter, energy and information. While I find hope and fascination with our techno-cultural evolution, many of my works express concern for ecological issues, which are often not considered within the realm of technological and cultural progress.

I have chosen interactive art in particular because it encourages active, self-determined relationships with a work of art and points to a co-evolved coupling between human, machine, nature and culture. The branching and joining of physical forms in my work echoes the behavioural flow and multiple directions an interactive piece may take in the act of self-organizing. I am compelled by open structures that define form but do not close the form off to the viewer. I use exposed electronics and mechanics as part of the aesthetic, in proposing structural relationships between wire, circuits and natural structures. I believe it is imperative that technological systems acknowledge and model the evolved wisdom of natural living systems, so they will inherently fuse, to permit an emergent and interdependent earth. Symbio-technoetic can describe this philosophy.

Autopoiesis is an artificial life robotic series of fifteen musical and robotic sculptures that interact with the public and modify their behaviours based on both the presences of the participants in the exhibition and the communication between each separate sculpture. [The] sculptures talk with each other through a hardwired network and audible telephone tones, which are a musical language for the group. *Autopoiesis* is 'self-making', a characteristic of all living systems. This characteristic of living systems was defined and refined by Francisco Varela and Humberto Maturana.

Autopoiesis [...] presents an interactive environment, which is immersive, detailed and able to evolve in real time by utilizing feedback and interaction from audience/participant members. The interactivity engages the viewer/participant, who in turn affects the system's evolution and emergence. This creates a system evolution as well as an overall group sculptural aesthetic. The structures themselves are constructed of cabernet sauvignon grape vines pulled into compression with steel wires. The joints are a custom-moulded urethane plastic. [...] The grape vines were selected to create an approachable natural sculpture that exists in the human biological realm.

Autopoiesis utilizes a number of unique approaches to create this complex and evolving environment. It uses smart sensor organization that senses the presence of the viewer/participant and allows the robotic sculpture to respond intelligently. [...] For example, at the top of each sculptural element (or arm), four passive infrared sensors [...] tell each arm to move in the direction of the viewer, while the active infra-red sensor located at the tip stops the arm as it arrives within inches of the viewer. This allows the sculpture to display both attraction and repulsion behaviours [and] give[s] the viewer a sense of the emotional state of the sculptural elements as they interact. Furthermore, in *Autopoiesis* the robotic sensors compare their sensor data through a central-state controller, so the viewer is able to walk through the sculptural installation and have the arms interact both individually and as a group. Some of the behaviours [include] 'follow the leader', where one arm is passing a 'mimic me' message to the next arm [and] 'flocking' behaviour, where they are all moving simultaneously, or flock out from the centre, where the arm in the centre sends a message for the other arms to follow. Higher and more rapid tones are associated with fear and the lower, more deliberate tonal sequences with relaxation and play. Other tones give the impression of the sculptures whistling to themselves. The telephone tones are a consistent language of intercommunication and manifest a sense of overall robotic group consciousness, where what is said by one, affects what is said by others. [...]

Autopoiesis continually evolves its own behaviours in response to the unique environment and viewer/participant inputs. This group consciousness of sculptural robots manifests a cybernetic ballet of experience, with the computer/machine and viewer/participant involved in a grand dance of one sensing and responding to the other.

Ken Rinaldo, extracts from 'Emergent Systems' and 'Autopoiesis' (2000). (http://kenrinaldo.com)

Benjamin Bogart and Philippe Pasquier
Context Machines//2013

Context Machines are generative artworks whose design is inspired by models of memory and creativity drawn from the cognitive sciences. In a traditional artistic context, the artist works directly in the material that is presented to the audience. In generative art, the artist manifests the concept in a system whose output is presented to the audience. This is a process of meta-creation: the building of systems that create media artefacts. Our development of *Context Machines* is manifest computationally and informed by cognitive models and theory, which are rarely exploited in generative art.

Our initial motivation leading to *Context Machines* is that their output be, to some degree, a surprise to us. Computational theories of complexity, emergence and non-determinism contribute to processes that enable surprising results. The creative behaviour of *Context Machines* is manifest in the generative representation presented to the audience. *Context Machines* are image-makers – but the process by which they generate images is more significant than the images themselves. Harold Cohen describes the significance of cognitive processes in image-making:

> An image is a reference to some aspect of the world which contains within its own structure and in terms of its own structure a reference to the act of cognition which generated it. It must say, not that the world is like this, but that it was recognized to have been like this by the image-maker, who leaves behind this record: not of the world, but of the act.[1]

Context Machines share a number of core features: they all involve a computer-controlled camera, used to collect images of their visual context, and use computational methods to generate novel representations. [...]

The artwork should relate itself to its context, without that relation being predetermined by the artist.

This is our central motivation and informs *Memory Association Machine*'s production and remains in the background of all *Context Machines*. The use of an 'intentional stance' frames the work as an autonomous entity that is capable of forming a relation to its context, which includes the audience. In order to form such a relation, the artwork must be embodied – albeit in a synthetic sense: the world impacts the system through the images collected by the machine, while the artwork impacts the world through the subtle effect of its representation on the viewer. For example, a rich and complex representation may encourage viewers to approach the work, which would increase the number of images of people collected by the system. In addition is the aspect of surprise, where the machine's representation should, to some degree, appear independent of the intention of the artist. This interest in surprise is analogous to the interest in erasing the 'artist's hand' in traditional art. In illusionistic painting, the lack of visible brushstrokes gives the viewer the impression that the work is magical and disconnected from the artist while simultaneously testifying to his or her skill. The creative behaviour of the *Context Machines* provides a similar magical quality: 'The signs of the will of a creator are sometimes less palpable in these objects than a manifestation of a "will" of their own.'[2] [...]

1 Harold Cohen, 'What is an Image?', in *Proceedings of the International Joint Conference on Artificial Intelligence* (Tokyo, 1979) 24.

2 [footnote 15 in source] Mitchell Whitelaw, *Metacreation: Art and Artificial Life* (Cambridge, Massachusetts: The MIT Press, 2004) 103.

Benjamin Bogart and Philippe Pasquier, extracts from '*Context Machines*: A Series of Situated and Self-Organizing Artworks', *Leonardo*, vol. 46, no. 2 (April 2013) 115–16.

UBERMORGEN.COM, with Paolo Cirio and Alessandro Ludovico
Google Will Eat Itself//2005

We generate money by serving Google text advertisements on our website GWEI. org. With this money we automatically buy Google shares. We buy Google via their own advertisement! Google eats itself – but in the end we'll own it! By establishing this model we deconstruct the new global advertisement mechanisms by rendering them into a surreal click-based economic model. We inject a social virus ('let's share their shares') into their commercial body, hidden under a polite and friendly graphic surface.

Then we hand over the common ownership of Google to the GTTP-community – Google to the People. A bit more in detail: One of Google's main revenue generators is the 'Adsense' program: it places hundreds of thousands of little

Google text-ads on websites around the world. Now we set up such an Adsense account for our GWEI.org website. Each time someone clicks on one of our Google text-ads, we receive a micro-payment and Google retains the same amount of money plus a certain percentage for its services – that's how they make their huge profit. Google pays us monthly by cheque. Each time we receive enough money, we buy the next Google share [NASDAQ: GOOG, currently trade between 150–250 USD]. This is the 'real new economy' – users get shares just for clicking! So how do we generate traffic and clicks? We use both a technical and a social level to reach our goal:

1. With a sophisticated and on-the-edge browser-server tool [flash/php] we generate a steady flow of clicks. We are locking the software on a limited amount of page views [~2500] and clicks [~200] per day. There is no difference between human clicking and this level of machine generated clicks – we are no script-kiddies but bastard artists.

2. Additionally to this we use our GTTP-community to spread the site and do page views and clicks. Low key social engineering, through our neural.it and UBERMORGEN.COM networks, can aggregate waves of inconspicuous clicks.

GWEI.org/gwei/ is the hidden website to showcase and unveil a total monopoly of information [Google search-engine + added services], a weakness of the new global advertisment system and the renaissance of the 'new economic bubble' – the 'reality' is that Google is currently valued more than all Swiss Banks together [sic]. Let's open their goldmine to the people as long as we are able to.

UBERMORGEN.COM (lizvlx and Hans Bernhard), with Paolo Cirio and Alessandro Ludovico, hand-out from exhibition, 'Deconstruction of Global D-Commerce', The Premises Gallery, Johannesburg, 2005. (http://gwei.org)

Mitchell Whitelaw
System Stories and Model Worlds//2005

[...] So far the discourse around software and generative art has focused largely on defining and contextualizing the field, and reflecting on its particular processes and materials – for example, the nature of 'code', or the question of software/process as art.

In order to come to grips with the works themselves, I would argue that any critique must be able to address the specifics of their generative systems: that the systems, not their outputs or residues, are the core of the work. System can be distinguished from code: code is the language-specific text that implements the abstract, formal structure that I will call system. [...]

How do we read such systems, critically? [...] Stefan Helmreich[1] and Katherine Hayles[2] have made strong analyses of a-life science, pursuing a basically deconstructive approach and arguing that a-life [artificial life] systems are fundamentally narrative in their operation. Moreover, for these critics a-life's narratives themselves 'reinscribe' particular assumptions about embodiment, subjectivity, gender, family and theology. These narratives are decoded in part from the discourse around the software system – Hayles, for example, makes use of a video representing Tom Ray's *Tierra* system, where Ray's biological and theological analogies are spelled out in the narration and the construction of the visualization. However, when Stefan Helmreich analyses John Holland's *Echo*, a platform for creating agent-based a-life simulations, he does so based on conversations with a programmer and inspection of the code; Helmreich's observations come as much from the defined formal structures of the software as they do from the discourse around those structures. These analyses suggest a way of reading systems as stories; they in turn create new, critical stories based on that interpretation.

So, a 'system story' is a translation or narration of the processual structures, ontology, entities and relations in a software system. Such stories are useful devices for opening up these systems to discussion and critique. System stories are not singular or objective; each one is a particular and situated reading. Nor are they floating signifiers though, since they draw on the concrete, formal object that is the software system. What generative art criticism needs are system stories that engage, in detail, with that formal object, and draw out its implications.

Hayles and Helmreich also provide an argument as to the importance of system stories. In their analyses, the narratives of artificial life are tacit, built-in assumptions which inform software models and simulations. In the case of a-life, there is an obvious relationship with the world 'outside' the simulation – with life as we know and live it. The critics warn us against mistaking these assumptions for 'the rules' of life – confusing the made with the given, or culture with nature. Similarly the value of system stories for generative art is in their ability to connect – critically, prospectively, speculatively – entities and relations within the system with entities and relations outside it. [...]

Simulation techniques are used in these works as generative devices, not as tools for modelling; but nonetheless the work is entirely shaped by the construction of its underlying system, its configuration of entities and relations. That configuration, what Brad Borevitz calls its 'logic' or 'systemacity',[3] is revealed to the user through a process of dynamic interaction; as Borevitz says, there is a kind of experiential reverse-engineering at play, as we map back from residue or output to system. Once again however, the system is core, and therefore surely the structure of that system is crucial. Especially in works using simulation and related techniques, abstract generative art performs *cosmogeny*: it brings forth a whole artificial world, saying, *here is my world*, and *here's how it works*. Once again, I will argue that this practice is in a unique position to explore and critique 'how it works'. Borevitz quotes Clement Greenberg on abstract painting and sculpture: 'like functional architecture and the machine, they *look* what they *do*.' So, what do they do? [...]

Engaged as it is in the pragmatics of generativity – of making something make something – generative software art turns to computationally expedient techniques. The simplest of these is combinatorics, or the playing out of permutations. Some recent visual generative art follows this approach, setting a simple system in motion and observing its outcomes. The results are visually complex, but the underlying system is surprisingly simple, as in some of the pieces in Casey Reas's *Software (Structures) #002* and *#003*, Jared Tarbell's *#003A* and *#003B*, and William Ngan's *#003B*.[4]

In this project the artist's focus was reflexive and processual: considering the 'natural language' specification of a structure and its varied implementation. Removed from that context, however, we are faced once again with the shape of the system, and the question of interpreting or responding to that configuration of entities and relations. The model worlds in these instances are pure machines, clockwork constellations. They transform determinism into aesthetic complexity, using scale of population and a kind of analytic or integrative visualization – displaying spatio-temporal relations rather than the entities themselves. [...]

Software (Structures) also shows examples of another common world-system, using techniques of physical simulation. Robert Hodgin's implementations of *#003*, and Ngan's *#003A*, both introduce simulations of momentum and gravity (disobeying the 'structure' in the process). Among the many other uses of this technique are Mark Napier[5] and Scott Snibbe's[6] works in the CODeDOC project.] [...] These techniques are pragmatic and effective, in generative terms: they create complex, dynamic interactions between elements, at a low computational cost. They also bring with them an immediate physical resonance, as we recognize these physical dynamics and infer the properties of the entities (their relative masses, the strength of gravity). [...]

Ngan writes of trying to imbue a 'sense of life' into the entities in his beautiful *#003A*; Hodgin describes the results as 'organic' and 'cellular'. Tarbell goes further, imagining the circle-entities 'experiencing' and 'choosing' intersections 'analogous to daily life'.[7] This critique is not intended to discourage or over-interpret these narratives, but rather to imagine the consequences of taking them more seriously, especially in their potential relationship with the 'outside' world.

This unfulfilled potential is especially clear in the way generative art uses multi-agent systems. In this ubiquitous technique, entities are explicitly defined and visualized, often literally traced as they move around a cosmos/canvas. Their relations with each other can be more complex than in a physical simulation, including 'flocking' behaviour, where individuals modify their motion based on that of their neighbours. [...]

Here too, the generative technique is effective in creating visual complexity, and emergent dynamic form; but again each multi-agent system encodes an ontology, a structure of entities and relations, which must be read as the core of the work. The entities themselves have characteristic properties: they are identical, or belong to a set of predefined types, and their properties and behaviour are static over time. The systems have a particular relation to time: they tend to be a series of instantaneous slices. The state of the system at one moment is a function of its state in the moment just passed (this is also true of physical simulations). In other words, history is all but absent. This is reflected in the construction of 'agent' and 'environment' in these systems. The environment here is (literally) a blank canvas, inert, empty space. Agents tend not to have a means of influencing that environment – even when they leave 'traces' in that space, the traces have no impact on the agents. The traces are visualization devices, not entities in the formal ontology. What kind of narrative is this? All these attributes can be explained as computationally pragmatic – the simplest or most efficient way to achieve the generative payoff of the swarm aesthetic. Again any referentiality of this system can be downplayed in favour of pure generative instrumentality. And again I would argue that in fact these works are fundamentally determined by this ontology, and that in a basic way we see it in the works (cf. Greenberg, above). The works visualize their structure of entities and relations. They model a world.

My concern is not for realism or to oppose the necessary abstraction that any simulation or agent-based system involves. Rather it is to point out that these systems encode, for whatever reason, specific ontologies, and that those ontologies in turn, especially in agent-based systems, present specific attributes: modes of being and relation, relationships between individual and group, morphology of groups, relations of individual and environment, models of being-in-time. Lev Manovich sees in such work an image of 'world as the dynamic networks of relations, oscillating between order and disorder – always vulnerable, ready to change with a single click of the user'.[8] This is true, the swarm aesthetic is enchanted with dynamic multiplicity, with shifting networks of relation, with coalescence and dispersal. But consider the subject or agent modelled here, if that's the story we want to tell: a clone in a crowd, unchanging, with no traction on the space it inhabits, existing in an ongoing, perpetual present. If these systems provide images of contemporary society then they are, at best, naive and utopian: a mass of identical (or typed) individuals, each contributing equally to the collective dynamic, each equally connected with and affecting all the others. As a social model this is a kind of idealized, frozen anarcho-democracy, where power relations (unequal causal connections) can never emerge.

This critique is simply a starting point; its flipside is more positive and important. If generative software art communicates system stories, particularly in the form of model worlds or ontologies, then it is potentially a platform for telling system stories that are more sophisticated, critical or experimental; it could take seriously the prospect that Manovich proposes, the potential of software and generative technique to provide images of, or rather *imaginations of*, the (social, cultural, personal, material ...) systems we live in. Generative art has a unique potential here, because unlike other art forms its basic materials are systems themselves. [...]

Casey Reas's works *Tissue* and *Microimage* begin to develop the homogeneous swarm, creating distinct 'species' of agent with distinctive (but again fixed) relationships. The added complexity of the interaction within the system is revealed in the images, as tangled clouds resolve into dark loops and braids. Similarly Ichitaro Masuda's recent work *Haohao*[9] has multiple species of agent, differentiated in size and colour, and attracted to and repelled from each other to varying (randomized) degrees. While Masuda's code reveals that the parameter for attraction is 'love', this is no agent-meets-agent story. Individuals form pseudo-stable clusters of five or more, where forces of attraction and repulsion are in equilibrium; these clusters might in turn orbit other groups, and are readily disrupted if another agent approaches. If there is a social story here, it is one of pursuit, desire and loss, but above all the delicate negotiation of local collectives or cliques. Once more this dynamic informs the aesthetic of the trail-paintings which the system produces, with tight gnarls and knots, as well as dense circular orbits and linear vectors.

These examples retain the usual disconnection between agent and environment – agents interact with each other, but have no functional impact on their world. However, this feature is not computationally or formally necessary, and in fact there seems to be a generative and aesthetic payoff for linking agent and environment more tightly. [...]

Narrative critiques reading software and generative art have a significant limitation, or rather a kind of grain or directionality. They can decompose a system, analyse the modes of being and relation that it encodes, but they have little to say about how those encodings play out, how they operate in a generative process. The emergence of complex, dynamic forms and behaviours from these local encodings is central to artists' interests in complex systems;[10] this is the moment of emergent generativity or the 'computational sublime'.[11] Once again, this is where generative art is in a unique and powerful position, in that unlike other forms of discourse, it can actually experiment with the emergent outcomes of particular ontologies, modes of being and relation. [...]

One of the further implications here is a reconsideration of the context for generative art. If it is fundamentally concerned with creating model ontologies, then we can imagine it in relation to other practices of formal modelling and simulation. These techniques have a long history in military strategy and geopolitics, but in recent years they have become more widespread. For example, a new branch of social science has emerged which uses simulation as a basic tool for testing 'explicit models of social phenomena'.[12] [...]

In other words, we are already being modelled, in artificial worlds that can fold back powerfully into the real. Like Helmreich, I would be very concerned if social modelling was used only to entrench our 'known features'. Unknown features must be more promising, and here again generative art can step in. Borevitz writes: 'If there is a chance that software will contribute significantly to a new politically relevant aesthetics, it lies in the way software shows us a way out of order, in and through order.'[13] Yes, but what's required is attention to the specifics of that order, its structures and properties. Generative art can, and must, do more than make images of complex systems; it can tinker critically with the systems themselves, then set them running: possible worlds.

If abstract or generative software art can, and sometimes does, work this way, where does this leave the binary of formalism/culturalism, or generative/software art? Perhaps the relation could be one of complementarity. 'Culturalist' software art [e.g. UBERMORGEN.COM's *Google Will Eat Itself*] has often focused on intervening critically, and practically, in existing software systems, reconfiguring them from the inside. In the process it shows up the latent cultural agency of software, but also its potential transformation. [...]

As I have argued, we can think of abstract software art, or generative art, as potentially exploring alternative modes of being and relation, telling stories but also literally toying with complex, dynamic systems, exploring them prospectively, and not (merely) as eye-candy machines, but as model worlds. To restate the binary: perhaps generative formalism [e.g. Reas's *Software (Structures)*] can be prospective and exploratory, where culturalism is more local, situated, concrete, interventionist. The two strands might in fact be complementary, and their critical potential might be far greater if we think them together, instead of apart.

1 [footnote 7 in source] Stefan Helmreich, *Silicon Second Nature: Culturing Artificial Life in a Digital*

World (Berkeley and Los Angeles: University of California Press, 1998) 163–4.

2 [8] N. Katherine Hayles, *How We Became Posthuman* (Chicago: University of Chicago Press, 1999) 227–31.

3 [9] Brad Borevitz, 'Super-Abstract: Software Art and a Redefinition of Abstraction', in *read_me Software Art & Cultures Edition*, ed. Olga Goriunova & Alexei Shulgin (Aarhus: Center for Digital AEstetik-forskning, 2004) 310–11.

4 [10] Casey Reas, et al., *Software (Structures)* (2004). (http://artport.whitney.org/com- missions/softwarestructures)

5 [11] Mark Napier, *3 dots*. (http://artport.whitney.org/commissions/codedoc/Napier/app/wave.html)

6 [12] Scott Snibbe, *Tripolar*. (http://artport.whitney.org/commissions/codedoc/Snibbe/Tripolar.html)

7 [13] William Ngan, Robert Hodgin and Jared Tarbell (2004). (http://artport.whit- ney.org/commissions/softwarestructures/s3_william_1/comments.html)

8 [4] Lev Manovich, 'Abstraction and Complexity' (2004). (http://www.manovich.net/DOCS/abstraction_complexity.doc)

9 [16] Ichitaro Masuda, *Haohao* (2004–5). (http://www.iamas.ac.jp/~madman03/haohao/)

10 [20] Casey Reas, 'Microimage' (2003). (http://reas.com/texts/microimage.html)

11 [21] Jon McCormack and Alan Dorin, 'Art, Emergence and the Computational Sublime', in *Proceedings of Second Iteration: A Conference on Generative Systems in the Electronic Arts*, ed. A. Dorin (Buenos Aires: CEMA, 2001) 67–81.

12 [24] Nigel Gilbert, 'Simulation: An Emergent Perspective' (1995–96). (http://cress.soc.surrey.ac.uk/resources/emergent.html)

13 [26] Brad Borevitz, 'Super-Abstract', op. cit., 311.

Mitchell Whitelaw, extracts from 'System Stories and Model Worlds', in *ReadMe 100*, ed. Olga Goriunova (Norderstedt, Germany: Herstellung und Verlag: Books on Demand GmbH, 2005) 139–52.

Geoff Cox
Generator: The Value of Software Art//2007

[...] [S]oftware art exemplifies process-orientated practice in a way that lends itself to critical work appropriate to contemporary conditions. [...] [O]lder definitions associated with generative art stress the formal rule-based and syntactical properties of software, and thus do not place sufficient emphasis on semantic concerns and social context. Although, in general, this may be the case, formal concerns are essential to understand the more cultural aspects and the generative or transformative aspects of software. The essay argues that taken together, the terms generative art and software art emphasize productive contradictions – inherent to both, and between the two. [...]

Inke Arns [...] stress[es] the distinction between earlier work using computers and software art, where the latter is '...not art that has been created with the help of a computer, but art that happens in the computer; software is not programmed by artists in order to produce autonomous artworks, but the software itself is the artwork. What is crucial here is not the result but the process triggered in the computer by the program code.'[1]

[...] But in the case of software, it is not simply a choice of process or product but of the interaction between source code and its executed form. [...] [The] privileging of execution, even if in combination with source code, avoids some of the contemporary practices associated with software art [e.g.] programs that are not necessarily executable, or executable only on a conceptual level (often referred to as 'codework'). Perhaps it is simply a case of generative art requiring improved description to shift emphasis from the object generated to the process of generation. [...]

[In the] exhibition 'CODeDOC', first for the Whitney Museum of American Art's 'artport' web site (2002), and later at Ars Electronica (2003), the curator Christiane Paul set the invited artist-programmers an instruction to 'connect and move three points in space' in a language of their choice (Java, C, Visual Basic, Lingo, Perl) and to exchange the code with the other artists for comments. [Here] code is taken to be part of the work and not simply meant to assist interpretation. [...]

The challenge for a critical practice in software art is to maintain contradiction in the process of transformation, for this is where politics is evident and where re-invention takes place. In [...] a contemporary situation where conceptual strategies have become the orthodoxy of contemporary art and effectively recuperated, radical art can be found in social energies not yet recognized as art. Perhaps software art and culture represents such an instance – for now at least.

1 [footnote 2 in source] Inke Arns, 'Read_Me, Run_Me, Execute_Me: Software and its Discontents, or It's the Performativity of Code, Stupid', in Olga Goriunova & Alexei Shulgin, eds, *Read_Me: Software Art & Cultures* (Århus: Digital Aesthetics Research Centre, 2004)184–5.

Geoff Cox, extracts from 'Generator: The Value of Software Art', in *Issues in Curating, Contemporary Art and Performance*, ed. Judith Rugg (Bristol: Intellect, 2007) 147–61.

WE CAN'T CONTROL SYSTEMS OR FIGURE THEM OUT. BUT WE CAN DANCE WITH THEM

Donella H. Meadows, Dancing with Systems, 2001

ENVIRONMENTAL AND SOCIAL SYSTEMS

R. Buckminster Fuller
Operating Manual for Spaceship Earth//1969

[S]ociety operates on the theory that specialization is the key to success, not realizing that specialization precludes comprehensive thinking. This means that the potentially-integratable-techno-economic advantages accruing to society from the myriad specializations are not comprehended integratively and therefore not realized, or they are realized only in negative ways. [...]

One of humanity's prime drives is to understand and be understood. All other living creatures are designed for highly specialized tasks. Man seems unique as the comprehensive comprehender and coordinator of local universe affairs. [...]

In organizing our grand strategy we must first discover where we are now; that is, what our present navigational position in the universal scheme of evolution is. To begin our position-fixing aboard our Spaceship Earth we must first acknowledge that the abundance of immediately consumable, obviously desirable or utterly essential resources have been sufficient until now to allow us to carry on despite our ignorance. Being eventually exhaustible and spoilable, they have been adequate only up to this critical moment. [...]

We begin by eschewing the role of specialists who deal only in parts. Becoming deliberately expansive instead of contractive, we ask, 'How do we think in terms of wholes?' If it is true that the bigger the thinking becomes the more lastingly effective it is, we must ask, 'How big can we think?' [...]

One of the modern tools of high intellectual advantage is the development of what is called general systems theory. Employing it, we begin to think of the largest and most comprehensive systems, and try to do so scientifically. We start by inventorying all the important, known variables that are operative in the problem. But if we don't really know how big 'big' is, we may not start big enough, and are thus likely to leave unknown but critical variables outside the system, which will continue to plague us. Interaction of the unknown variables inside and outside the arbitrarily chosen limits of the system are probably going to generate misleading or outrightly wrong answers. If we are to be effective, we are going to have to think in both the biggest and most minutely-incisive ways permitted by intellect, and by the information thus far won through experience. [...]

Can we think of, and state adequately and incisively, what we mean by universe? For universe is, inferentially, the biggest system. If we could start with universe, we would automatically avoid leaving out any strategically critical variables. [...]

Holding to the scientists' experiences as all important, I define universe, including both the physical and metaphysical, as follows: The universe is the aggregate of all of humanity's consciously-apprehended and communicated experience with the non-simultaneous, non-identical and only partially overlapping, always complementary, weighable and unweighable, ever omni-transforming, event sequences. [...]

Having adequately defined the whole system, we may proceed to subdivide progressively. This is accomplished through progressive division into two parts, one of which, by definition, could not contain the answer – and discarding of the sterile part. Each progressively-retained live part is called a 'bit', because of its being produced by the progressive binary 'yes' or 'no' bi-section of the previously residual live part. The magnitude of such weeding operations is determined by the number of successive bits necessary to isolate the answer. [...]

How many 'bisecting bits' does it take to get rid of all the irrelevancies and leave in lucid isolation that specific information you are seeking? We find that the first subdividing of the concept of universe-bit one is into what we call a system. A system subdivides universe into all the universe outside the system (macrocosm) and all the rest of the universe which is inside the system (microcosm), with the exception of the minor fraction of universe which constitutes the system itself. The system divides universe not only into macrocosm and microcosm but also coincidentally into typical conceptual and non-conceptual aspects of universe – that is, an overlappingly-associable consideration, on the one hand, and, on the other, all the non-associable, non-overlappingly-considerable, non-simultaneously-transforming events of non-synchronizable disparate wave frequency rate ranges. [...]

Synergy is the only word in our language that means behaviour of whole systems, unpredicted by the separately observed behaviours of any of the system's separate parts or any sub-assembly of the system's parts. [...]

There is nothing about an electron alone that forecasts the proton, nor is there anything about the Earth or the Moon that forecasts the coexistence of the Sun. The solar system is synergetic – unpredicted by its separate parts. But the interplay of Sun as supply ship of Earth and the Moon's gravitationally produced tidal pulsations on Earth all interact to produce the biosphere's chemical conditions, which permit but do not cause the regeneration of life on Spaceship Earth. This is all synergetic. There is nothing about the gases given off respiratorily by Earth's green vegetation that predicts that those gases will be essential to the life support of all mammals aboard Spaceship Earth, and nothing about the mammals that predicts that the gases which they give off respiratorily are essential to the support of the vegetation aboard our Spaceship Earth. Universe is synergetic. Life is synergetic. [...]

R. Buckminster Fuller, extracts from *Operating Manual for Spaceship Earth* (Carbondale, Illinois: Southern Illinois University Press, 1969); Pocket Book edition (New York: Simon & Schuster, 1970) 12–13; 51–66.

James Lovelock
Geophysiology: The Science of Gaia//1989

There is growing recognition of the inadequacy of the separated disciplinary approach for the solution of planetary scale problems. To understand even the atmosphere, which is the simplest of the planetary compartments, it is not enough to be a geophysicist; knowledge of chemistry and biology is also needed. It might seem that research teams that include experts in each of the different disciplines would resolve the problem, but anyone who has attended gatherings of experts knows that each expert speaks but does not or cannot listen. What might help would be a broader-based general science, or a scientific operating system, that provides an environment within which the separate disciplines could interact.

Contemporary concerns have developed from the consequences of changes made by humans in the composition of the atmosphere and the nature of the land surface and biota. In many ways these modern concerns echo similar concerns about the human body early in the development of medicine. In the late nineteenth century the sciences of biochemistry and microbiology were well advanced but largely disconnected and not very helpful to those with disease. Advances in medicine were, however, vastly enabled by the existence of the general science of physiology. This science was transdisciplinary and also recognized the essentially emergent properties of a living organism. If one is interested in how our core temperatures are maintained at 37 degrees centigrade, a biochemical approach to a solution of the problem is fruitless. Temperature regulation is a systems control problem. But by starting with physiology, the biochemical aspects involving, for example, oxidative metabolism naturally fit into place. The main purpose of this paper will be to put forward an analogous Earth science, geophysiology, as the transdisciplinary environment for planetary scale problems, particularly those involving a wide range of disciplines. Where it is postulated, even though not proved, that emergent properties exist, it may be useful for practical purposes to consider the Earth as if it were a living organism.

Before the nineteenth century, scientists were comfortable with the notion of a living Earth. One of them was James Hutton, who has often been called the father of geology. Hutton ['Theory of the Earth', 1788] likened the Earth to a superorganism and recommended physiology as the science for its investigation. [...]

Hutton's wholesome view of the Earth was discarded early in the last century. I think that this may have been a consequence of a growing interest in origins and in evolutionary theories both for Earth and for life sciences. For biologists there was Darwin's great vision of the evolution of the species of organisms by natural selection.

For the geologists there was the wholly independent theory that the evolution of the material environment was simply a matter of chemical and physical determination. The divorce of the Earth and life sciences in the nineteenth century was inevitable. There was a rapid increase in the supply of information about the Earth as exploration and exploitation developed. But the techniques for looking at organisms were very different from those for looking at the ocean, the air and the rocks. It must have been an exciting period of science. There were few inclined to stand back and take a broader view or try to keep alive Hutton's superorganism. What is remarkable is not the division of the sciences, but that two distinct and very different theories of evolution could coexist even until today.

The reason for endurance of the division is, I think, a mutual acceptance by Earth and life scientists of the anaesthetic notion of adaption. [...]

Adaptation is a dubious notion, for in the real world the environment, to which the organisms are adapting, is determined by their neighbours' activities rather than by the blind forces of chemistry and physics alone. In such a world, changing the environment could be part of the game, and it would be absurd to suppose that organisms would refrain from changing their material environment, if by so doing they left more progeny. In his time, of course, Darwin did not know, as we do now, that the air we breathe, the oceans and the rocks are all either the direct products of living organisms or have been greatly modified by their presence. In no way do organisms just 'adapt' to a dead world determined by physics and chemistry alone. They live with a world that is the breath and bones of their ancestors and that they are now sustaining. [...]

Like co-evolution, Gaia reflects the apartheid of Victorian biology and geology, but it goes much further. Gaia theory is about the evolution of a tightly coupled system whose constituents are the biota and their material environment, which comprises the atmosphere, the oceans and the surface rocks. Self-regulation of important properties, such as climate and chemical composition, is seen as a consequence of this evolutionary process. Like living organisms and many closed loop self-regulating systems, it would be expected to show emergent properties; that is, the whole will be more than the sum of the parts. This kind of system is notoriously difficult, if not impossible, to explain by cause and effect logic, as practising inventors know to their cost. [...]

Engineers and physiologists have long been aware of the subtleties of feedback. Homeostasis is only possible when feedback is applied at the right amplitude and phase and when the system's time constants are appropriate. Both positive and negative feedback can lead to stability or instability, depending on the timing of their application. Theoretical ecology models, notorious for their intractable mathematics, would not surprise an engineer, who would see them in his words as 'open loop systems' where feedback was applied, or happened by chance, in an arbitrary manner. By contrast, geophysiological models, such as *Daisyworld* [a computer simulation of a hypothetical world in orbit], include feedback, negative and positive, in a coherent manner. As a consequence, the models are robust and stable and will happily accommodate any number of non-linear equations and still prefer to relate with stable attractors. [...]

I do not disagree with those who propose that some, or even a large proportion, of the total regulation of any chosen Earth property can be explained by deterministic chemistry and physics. Living systems use chemistry economically. They do not strive ostentatiously to do better than blind chemistry or physics because there is no need. The purpose of Gaia is to offer a new way of looking at the Earth and to make predictions that can be tested experimentally. Had it not been for the curiosity stimulated by thoughts on the mechanisms of Gaia, none of the important trace gases dimethyl sulfide, carbon disulfide, methyl iodide and chloride would have been sought and found when they were [...]. To conclude, Gaia theory provokes us to think about three things:

1. Life is a planetary scale phenomenon. There cannot be sparse life on a planet. It would be as unstable as half of an animal. Living organisms have to regulate their planet; otherwise, the ineluctable forces of physical and chemical evolution would render it uninhabitable.

2. Gaia theory adds to Darwin's great vision. There is no longer any need to consider the evolution of the species separately from the evolution of their environment. The two processes are tightly coupled as a single indivisible process. It is not enough merely to say that the organism that leaves the most progeny succeeds. Success also depends upon coherent coupling between the evolution of the organism and the evolution of its material environment.

3. Lastly, it may turn out that the gift of Gaia to geophysics is the reduction of Alfred Lotka's insight [*Elements of Physical Biology*, 1925] to practice: a way to look at the Earth mathematically that joyfully accepts the non-linearity of nature without being overwhelmed by the limitations imposed by the chaos of complex dynamics.

James Lovelock, extracts from 'Geophysiology: The Science of Gaia', *Reviews of Geophysics*, no. 17 (May 1989) 215–22 [footnotes not included].

Helen Mayer Harrison and Newton Harrison
Shifting Positions towards the Earth: Art and Environmental Awareness//1993

Our work begins when we perceive an anomaly in the environment that is the result of opposing beliefs or contradictory metaphors. Moments when reality no longer appears seamless and the cost of belief has become outrageous offer the opportunity to create new spaces – first in the mind and thereafter in everyday life.

We understand the universe as a giant conversation taking place simultaneously in trillions of voices and billions of languages, most of which we could not conceive of even if we knew that they existed. Of those voices whose existence has impinged on our own to the degree that we can become aware of them, we realize that our awareness is imperfect at best. Therefore, it seems to us that the casual and wanton destruction and disruption of living systems of whose relationships we know so little requires extraordinary hubris.

For us, everything started with a decision made in the late 1960s to deal exclusively with issues of survival as best we could perceive them. Each body of work sought a larger or more comprehensive framing or understanding of what such a notion might mean and how we, as artists, might express it. For example, in *The Seventh Lagoon* of *The Lagoon Cycle*, we came up with the statement: 'but that would require reorienting consciousness around a different database'.

We are now exploring what such a statement might mean – unpacking our intuitive ideas. Our most recent work opens up the idea of setting up an eco-security system, a safety net for the ecology not unlike a social security system. However, there are issues such as the population explosion that need a separate and comprehensive address, for just as prairie grass would displace everything that is not itself, so would any expanding population. The notion that ingenious technology will resolve population pressures on the one hand and generate infinitely expanding markets on the other is simply an illusion. It is too easy to forget that every entrepreneurial act, even recycling, is itself a tax on the ecosystem. [...]

The Lagoon Cycle (1973–85)

This work is, in part, a mural 360 feet long, averaging 8 feet tall, in sixty parts. It was completed over the period 1973 to 1985. It is portable, done on photomural paper mounted on heavy cotton duck. The materials are photography, oil, graphite, crayon and ink. It was first exhibited in complete form at the Johnson Gallery at Cornell University in 1985 and then later at the Los Angeles County Museum of Art.

The Lagoon Cycle can be read as a story in seven parts; each part, as in a picaresque novel, is its own story. It can be read as an array of storyboards for a very unusual movie. As artists, we see it as an environmental narrative, one of whose properties is to envelop the viewer with its form and subject matter. For us, this work relates to other twentieth-century environmental works as well as to the myriad mural programs of the past.

The Lagoon Cycle unfolds as a discourse between two characters who discuss the ways in which the metaphors we live by affect what we do to each other and to the environment. It casts light on how we create our world view and are in turn created by it. *The Lagoon Cycle* is named after the estuarial lagoons that are endangered everywhere; the lagoons are used as a metaphor for culture and even for life itself.

The story concerns two characters who begin a search for a 'hardy creature who can live under museum conditions' and who are transformed by this search. The characters define themselves in *The First Lagoon* by the differences in their values and perceptions, with one naming himself Lagoon Maker and the other naming herself Witness. Both proceed to live up to their names, although they finally surrender them as circumstances push the two characters into constructing ever-larger frames for their discourse.

The Sixth Lagoon: On Metaphor and Discourse

The Fifth Lagoon deals with the Salton Sea, which was formed by flood flow released by human error from the canals along the Colorado River. *The Sixth Lagoon* treats the entire Colorado River basin. Lagoon Maker and Witness reflect on the insights they have gained through observing aquatic systems. They expand the scale of their thinking from the Salton Sea to the Colorado River watershed, which has been changed by lifestyles that demand vast amounts of electricity and irrigation. The exploding megatechnology of the twentieth century has shocked the environment and does not have time to 'niche itself in'. Witness sees all nature as a discourse between the elements, and both characters urge, 'Pay attention to the discourse between belief systems and environmental systems.'

Pay attention to the flow of waters
Pay attention to the integrity of the waters flowing
Pay attention to where the waters are flowing
Pay attention to where the waters desire to flow
Pay attention to where the waters are willed to flow
Pay attention to the flow of waters and the mixing of salts
Pay attention to the flow of waters and the mixing with earth
Attend to the integrity of the discourse between earth and water
the watershed is an outcome

Pay attention to the discourse between earth water and men
interruption is an outcome
Pay attention to the meaning of the nature of such discourse and the nature of the meaning of interruption
After all a discourse is a fragile transitory form an improvisation of sorts
And anyone may divert a discourse of any kind into another direction if they do not value its present state
Pay attention to changes of state [...]

As we stated earlier, we believe that the universe is a giant conversation and that any introduction of new ideas, new metaphors, or new possibilities can change that conversation. Although we have built works, we think that changes in the conversation that lead to attitudinal and behavioural changes are as significant as any 'built' work.

Helen Mayer Harrison and Newton Harrison, extracts from 'Shifting Positions towards the Earth: Art and Environmental Awareness', *Leonardo*, vol. 26, no. 5, special issue: *Art and Social Consciousness* (1993) 160–79.

David Dunn and Jim Crutchfield
Entomogenic Climate Change: Insect Bio-Acoustics and Future Forest Ecology//2009

Forest ecosystems result from a dynamic balance of soil, plants, insects, animals and climate. The balance, though, can be destabilized by outbreaks of tree-eating insects. These outbreaks in turn are sensitive to climate, which controls precipitation. Drought stresses trees, rendering them vulnerable to insect predation. The net result is increased deforestation driven by insects and modulated by climate.

For their part, many species of predating insects persist only to the extent that they successfully reproduce by consuming and living within trees. Drought-stressed trees are easier to infest compared to healthy trees, which have more robust defences against attack. To find trees suitable for reproduction, insects track relevant environmental indicators, including chemical signals and, probably, bio-acoustic ones emitted by stressed trees. At the level of insect populations, infestation dynamics are sensitive to climate via seasonal temperatures. Specifically, insect populations increase markedly each year in which winters are short and freezes less severe. The net result is rapidly changing insect populations whose dynamics are modulated by climate.

Thus, via temperature and precipitation, climate sets the context for tree growth and insect reproduction and also for the interaction between trees and insects. At the largest scale, climate is driven by absorbed solar energy and controlled by relative fractions of atmospheric gases. The amount of absorbed solar energy is determined by cloud and ground cover. Forests are a prime example, as an important ground cover that absorbs, uses and re-radiates solar energy in various forms. At the same time forests are key moderators of atmospheric gases. Trees expel oxygen and take up carbon dioxide in a process that sequesters carbon from the atmosphere in solid form. As plants and trees evolved, in fact, they altered the atmosphere sufficiently that earth's climate, once inhospitable, changed and now supports a wide diversity of life.

There are at least three stories here: those of the trees, the insects and the climate. They necessarily overlap, since the phenomena and interactions they describe co-occur in space and in time. Their overlap hints at an astoundingly complicated system, consisting of many cooperating and competing components; the health of any one depends on the health of others. [...] How are we to understand the individual views as part of a larger whole? In particular, what can result from interactions between the different scales over which insects, trees and climate adapt?

Taking the stories together, we have, in engineering parlance, a feedback loop: Going from small to large scale, one sees that insects reproduce by feeding on trees; forests modulate insect reproduction, and precipitation controls tree growth. The feedback loop of insects, trees and climate means that new kinds of behaviour can appear – dynamics caused not by a single player but by their interactions. Importantly, such feedback loops can maintain ecosystem stability or lead to instability that amplifies even small effects to the large scale.

Here we give a concrete example of the dynamic interaction between insects, trees and climate. We focus on the role that bark beetles (Scolytidae or, more recently, Curculionidae: Scolytinae) play in large-scale deforestation and consequently in climate change. Bark beetles are emblematic of many different insect species that now participate in rapid deforestation. Likewise, we primarily focus on the North American boreal forests because of their unique characteristics but also as representative of the vulnerability of all types of forest ecosystems. Thus, the picture we paint here is necessarily incomplete; nonetheless, these cases serve to illustrate the complex of interactions implicated in the feedback loop and also the current limits to human response.

Although they are not alone, bark beetles appear to be an example of a novel player in climate change. Unlike the climatic role that inanimate greenhouse gases are predicted to play in increasing global temperature over the next century, bark beetles represent a biotic agent that actively adapts on the shorter time scale of years but still can cause effects, such as deforestation, at large spatial scales. To emphasize the specificity and possible autonomy of this kind of biological, non-human agent, we refer to the result as entomogenic climate change.

A detailed analysis of the problem of entomogenic climate change leads us to make a number of constructive suggestions for increased attention to relatively less familiar domains of study, including micro-ecological symbiosis and its non-linear population dynamics, and insect social organization. Here we emphasize in particular the role that bark beetle bio-acoustic behaviour must have in their evolving multiple survival adaptations, which, it appears, fills in significant gaps in the explanatory model of infestation dynamics. One goal is to stimulate interdisciplinary research appropriate to the complex of interactions implicated in deforestation and to discovering effective control strategies.

Forest Health and Climate

The Earth's three great forest ecosystems – tropical, temperate and boreal – are of irreplaceable importance to its self-regulating balance. Their trees help to regulate the Earth's climate, provide essential timber resources and create a diversity of habitat and nutrients that support other forms of life, including millions of people. Forests contribute to global climate dynamics through a carbon cycle in which atmospheric carbon dioxide is converted into an immense carbon pool. [...]

All forms of deforestation, human and natural, directly impact climatic conditions by attenuating or delaying the carbon cycle. In concert with well-documented greenhouse gas effects that drive global atmospheric change, the potential loss of large areas of these forests, combined with accelerating deforestation of tropical and temperate regions, may have significant future climate impacts beyond already dire predictions. Ice core studies reveal that the Earth's climate has varied cyclically over the past 450,000 years. Temperatures have been closely tied to variations in atmospheric carbon dioxide, in a cycle that recurs on the time scale of millennia. Vegetation has been forced to adapt. The boreal forests are, in fact, highly vulnerable to these climate shifts. Examination of fossil pollen and other fossil records shows that, in response to temperature variations over the past millennia, North American boreal forests have changed radically many times. The unique sensitivity of these forests' tree species to temperature suggests that the predicted warmer climate will cause their ecological niches to shift north faster than the forests can migrate.

One major consequence of boreal deforestation is increasing fire risk. Over the next half-century, the Siberian and Canadian boreal forests will most likely see as much as a 50 percent increase in burnt trees. One of the major sources fuelling these fires will be dead and dying trees killed by various opportunistic insect species and their associated micro-organisms.

Paralleling concerns about the boreal forests, in recent years there has been a growing awareness of extensive insect outbreaks in various regional forests throughout the western United States. As consecutive summers of unprecedented forest fires consumed the dead and dying trees, a new concern emerged: insect-driven deforestation is a threat connected to global climate change. In fact, climate experts, forestry personnel and biologists have all observed that these outbreaks are an inevitable consequence of a climatic shift to warmer temperatures. [...]

It is now well established that mountain pine beetles have slipped through mountain passes from the Peace River country in northern British Columbia to Alberta, the most direct corridor to the boreal forests. If the beetle is successful at adapting to and colonizing Canada's jack pine, there will be little to stop it moving through the immense contiguous boreal forest, all the way to Labrador and the North American East Coast. It then will have a path down into the forests of eastern Texas. Entomologist Jesse Logan describes this as 'a potential geographic event of continental scale with unknown, but potentially devastating, ecological consequences'.

Continental migration aside, if the beetles infest the high-elevation conifers, the so-called five-needle pines, of the western United States, this will reduce the snow-fence effect that these alpine forests provide. Snow fences hold windrows of captured snow that are crucial to the seasonal conservation and distribution of water from the Rocky Mountains. This is one of the primary origins of the water that sources several major river systems in North America. Every western state is contending with various rates of unprecedented insect infestation not only by many different species of Scolytidae but also by other plant-eating insects.

These and other rising populations of phytophagous insects are now becoming recognized as a global problem and one of the most obvious and rapidly emerging consequences of global climate change. Over the past fifteen years, there have been reports of unusual and unprecedented outbreaks occurring on nearly every continent. [...]

The Bio-acoustic Ecology Hypothesis

One of the more under-appreciated research domains regarding bark beetles concerns their remarkable bio-acoustic abilities. [...]

Past research suggested that sound-making and perception in bark beetles was secondary to their use of chemical-signalling mechanisms. Most studies addressing acoustic behaviour concentrated on sound generation, and only in its relationship to chemical signalling. These include the role stridulation sound-making has in controlling attack spacing between entry points in the host or in the triggering of pheromone release between genders. The resulting view is that bark beetles use a combination of chemical and acoustic signals to regulate aggression, attack on host trees, courtship, mating behaviour and population density.

An emphasis on pheromone-based communication may very well have led to a lack of follow-up on the possibility that host trees themselves produce acoustic cues that attract pioneer beetles. Perhaps the earliest proposal dates to 1987, when William Mattson and Robert Haack (of the USDA and Forest Service, respectively) speculated that cavitation events in trees might produce acoustic signals audible to plant-eating insects. Cavitation occurs in trees through breaking of water columns conducting the xylem tissue of leaves, stems and trunks. The assumption has been that the sounds are vibrations coming from individual cells collapsing, which is due to gradual dehydration and prolonged water stress. While cavitation produces some acoustic emissions in the audible range (20 Hz – 20 kHz), most occur in the ultrasound range (20 – 200 kHz and above). [...]

Recent fieldwork by Dunn focused on sound production by the pinion engraver beetle (Ips confusus). Sounds were recorded within the interior phloem layer of pinion trees, often adjacent to beetle nuptial chambers. A rich and varied acoustic ecology was documented – an ecology that goes beyond the previously held assumptions about the role of sound within this species. Another important observation was that much of the sound production by this species has a very strong ultrasonic component. Since communication systems seldom evolve through investing substantial resources into portions of the frequency spectrum that an organism cannot both generate and perceive, this raised the question of whether or not bark beetles have a complementary ultrasonic auditory capability. Recent laboratory investigations by Carleton University biologist Jayne Yack have also revealed ultrasound components in some bark beetle signals and indirect evidence that beetles possess sensory organs for hearing airborne sounds.

One possible implication that arises from the combination of these laboratory and field observations is that various bark beetle species may possess organs capable of hearing ultrasound for con-specific communication. If so, these species would be pre-adapted for listening to diverse auditory cues from trees.

This in turn raises an important issue not addressed by previous bark beetle bio-acoustic research. A very diverse range of sound signalling persists well after the putatively associated behaviours – host selection, coordination of attack, courtship, territorial competition and nuptial chamber excavations – have all taken place. In fully colonized trees the stridulations, chirps and clicks can go on continuously for days and weeks, long after most of the associated behaviours will have apparently run their course. These observations suggest that these insects have a more sophisticated social organization than previously suspected – one that requires ongoing communication through sound and substrate vibration.

The above acoustic fieldwork led us to conclude that there must be a larger range of forms of insect sociality and, therefore, means of organizational communication. More precise understanding of these forms of social organization may improve our ability to design control systems, whether these are chemical, acoustic or biological.

Closing the Loop

The eventual impact that insect-driven deforestation and global climate change will have on the Earth's remaining forests ultimately depends on the rate at which temperatures increase. The impacts will be minimized if that rate is gradual, but increasingly disruptive if the change is abrupt. Unfortunately, most climate change projections now show that a rapid temperature increase is more likely. [...]

One conclusion appears certain. Extensive deforestation by insects will convert the essential carbon pool provided by the Earth's forests into atmospheric carbon dioxide. Concommitently, the generation of atmospheric oxygen and sequestration of carbon by trees will decrease.

Most immediately, though, as millions of trees die, they not only cease to participate in the global carbon cycle but become potential fuel for more frequent and increasingly large-scale fire outbreaks. These fires will release further carbon dioxide into the atmosphere and do so more rapidly than the natural cycle of biomass decay. The interactions between these various components and their net effect are complicated at best – a theme running throughout the entire feedback loop. [...]

The repeated lesson of complex, non-linear dynamical systems, though, is that the apparent stability of any part can be destabilized by its place in a larger system. [...]

Taken alone, the potential loss of forests is of substantial concern to humans.

When viewing this system as a feedback loop, however, the concern is that the individual components will become part of an accelerating positive feedback loop of sudden climatic change. Such entomogenic change, given the adaptive population dynamics of a key player (insects), may happen on a very short timescale. This necessitates a shift in the current characterization of increasing insect populations as merely symptomatic of global climate change to a concern for insects as a significant generative agent.

In addition to concerted research in bio-acoustics, micro-ecological symbiosis and dynamics, and insect social organizations, these areas, in conjunction with the field of chemical ecology, must be integrated into a broader view of multi-scale population, evolutionary and climate dynamics. In this sense, the birth of chemical ecology serves as an inspiration. It grew out of an interdisciplinary collaboration between biology and chemistry. It is precisely this kind of intentional co-operation between disciplines – but over a greater range of scales – that will most likely lead to new strategies for monitoring and defence against what seems to be a growing threat to the world's forests, and ultimately to humanity itself.

David Dunn and Jim Crutchfield, extracts from 'Entomogenic Climate Change: Insect Bio-Acoustics and Future Forest Ecology', *Leonardo*, vol. 42, no. 3 (June 2009) 239–44 [footnotes not included].

Pierre Bourdieu
The Field of Cultural Production, or The Economic World Reversed//1983

[...] Few areas more clearly demonstrate the heuristic efficacy of *relational* thinking than that of art and literature. Constructing an object such as the literary field requires and enables us to make a radical break with the [conventional approach] which tends to foreground the individual, or the visible interactions between individuals, at the expense of the structural relations – invisible, or visible only through their effects – between social positions that are both occupied and manipulated by social agents, which may be isolated individuals, groups or institutions. [...]

The literary or artistic field is a *field of forces*, but it is also a *field of struggles* tending to transform or conserve this field of forces. The network of objective relations between different positions implement in their struggles to defend or improve their positions (i.e. their position-takings), strategies which depend for their force and form on the position each agent occupies in the power relations (*rapports de force*).

Every position-taking is defined in relation to the *space of possibles* which is objectively realized as a *problematic* in the form of the actual or potential position-taking corresponding to the different positions; and it receives its distinctive *value* from its negative relationship with the coexistent position-takings to which it is objectively related and which determine it by delimiting it. It follows from this, for example, that a *prise de position* changes, even when it remains identical, whenever there is change in the universe of options that are simultaneously offered for producers and consumers to choose from. The meaning of a work (artistic, literary, philosophical, etc.) changes automatically with each change in the field within which it is situated for the spectator or reader. [...]

When a new literary or artistic group makes its presence felt in the field of literary or artistic production, the whole problem is transformed, since its coming into being, i.e. into difference, modifies and displaces the universe of possible options; the previously dominant productions may, for example, be pushed into the status of outmoded (*déclassé*) or classic works.[…]

The work of art is an object which exists as such only by virtue of the (collective) belief which knows and acknowledges it as a work of art. Consequently, […] a rigorous science of art must […] take into account everything which helps to constitute the work as such, not least the discourses of direct or disguised celebration which are among the social conditions of production of the work of art *qua* object of belief.

The production of discourse (critical, historical. etc.) about the work of art is one of the conditions of production of the work. Every critical affirmation contains, on the one hand, a recognition of the value of the work which occasions it, which is thus designated as worthy object of legitimate discourse […] and on the other hand an affirmation of its own legitimacy. Every critic declares not only his judgement of the work but also his claim to the right to talk about it and judge it. In short, he takes part in a struggle for the monopoly of legitimate discourse about the work of art, and consequently in the production of the value of the work of art. (And one's only hope of producing scientific knowledge – rather than weapons to advance a particular class of specific interests – is to make explicit to oneself one's position in the sub-field of the producers of discourse about art and the contribution of this field to the very existence of the object of study.) […]

[T]he sociology of art and literature has to take as its object not only the material production but also the symbolic production of the work […]. It therefore has to consider […] the producers of the meaning and value of the work – critics, publishers, gallery directors, and the whole set of agents whose combined efforts produce consumers capable of knowing and recognizing the work of art as such. […]

The literary or artistic field is at all times the site of a struggle between the two principles of hierarchization: the heteronomous principle, favourable to those who dominate the field economically and politically (e.g. 'bourgeois art') and the autonomous principle (e.g. 'art for art's sake'), which those of its advocates who are least endowed with specific capital tend to identify with degree of independence from the economy, seeing temporal failure as a sign of election and success as a sign of compromise. […]

[In t]he struggle […] over the imposition of the legitimate mode of cultural production […], the artists and writers who are richest in specific capital [recognition] and most concerned for their autonomy are considerably weakened by the fact that some of their competitors identify their interests with the dominant principles of hierarchization and seek to impose them even within the field, with the support of the temporal powers. [...] In the struggle to impose the legitimate definition of art and literature, the most autonomous producers naturally tend to exclude 'bourgeois' writers and artists, whom they see as 'enemy agents'. [...]

The [...] definition of the writer (or artist, etc.) is an issue at stake in struggles in every literary (or artistic, etc.) field. In other words, the field of cultural production is the site of struggles in which what is at stake is the power to impose the dominant definition of the writer and therefore to delimit the population of those entitled to take part in the struggle to define the writer [or artist]. [...]

Pierre Bourdieu, extracts from 'The Field of Cultural Production, or The Economic World Reversed', *Poetics*, vol. 12, no. 4–5 (November 1983) 311–24 [footnotes not included].

Nick Prior
Putting a Glitch in the Field: Bourdieu, Actor Network Theory and Contemporary Music//2008

[...] Glitch has become an influential presence in music since the late 1990s and has gained credence as a contemporary form of sonic experimentation based on computer-generated clusters of rhythmic pulses, skips, clicks and scratches. Its development out of commercially-restricted scenes into more mainstream musical environments follows a logic present in Pierre Bourdieu's analysis of the chiasmatic structure of cultural fields, where the position-taking of artists is meaningful only in relation to a dynamic space of social relations governed by the twin poles of economic and cultural capital.[1]

In this sense, it will be argued, we learn a lot about the social dynamics of stylistic practice from greater sensitization to its position in a structured setting of socio-economic relations partly defined by the social characteristics and position-takings of the musicians themselves. Bourdieu's cultural sociology pits itself very effectively against aesthetic writings on glitch, precisely because it refuses to cut analysis off at the stylistic boundaries of the work. But there are some outstanding questions, one of which is addressed towards the end of this article. In a context that cries out for attention to a range of agents involved in cultural production, to what extent is there room for a sufficiently complex treatment of technology under Bourdieu's corpus of ideas? [...] One possible supplementary position comes from actor network theory, a theory that treats the realm of technologies as bound to the human world in ways other than that of instruments, tools or social weapons. When technology is considered a true form of mediation, it will be argued, it is possible to extend the range of objects in Bourdieu's fields to include those devices, techniques and artefacts that permit the solidification and transformation of field relations. Despite some conceptual incommensurability, rubbing these two traditions against each other sheds light on the complex human/non-human entanglements and field trajectories of contemporary styles such as glitch. [...]

Bourdieu's concept of field [...] provides a heuristic for the analysis of 'historically constituted areas of activity with their specific institutions and their own laws of functioning'.[2] These fields consist of sets of historical relations between positions grounded in specific forms of power or capital. Conceptually, the field is an immediate invitation to think *relationally* about the actions of social agents who, propelled by their habituses, compete for particular values specific to that field.[3] It is the interactions and conflicts between these agents over the prizes available that define the precise contours of the field, particularly the limits of what is found to be acceptable as the stakes in the field. In which case, the field is also a space of competition, the analogy being a game of chess where players enter the game and position themselves according to the powers and moves available to them. [...]

In Bourdieu's hands, then, the field becomes a network of objective relations between agents, but also larger groupings and institutions distributed within a space of possible positions. Its function is not merely to describe a logic of struggle between agents, but also a grander attempt to examine how modern societies are themselves defined by an architecture of overlapping spheres such as artistic fields, economic fields and scientific fields. In fact, it is the relationship that particular fields have to what Bourdieu calls the 'field of power', the broader political field, that defines their ability to resist the penetrations of outside forces such as the market. In the case of the cultural field, autonomy is dependent on the increasingly dualistic structure of a space defined by two logics of capital, economic and cultural. It is these species of capital that internally divide the cultural field into two sub-fields: on the one hand, the 'de-limited' sub-field of production and, on the other, the 'large-scale' or 'heteronomous' field of production. While the delimited field is defined by its distance from commercial mass markets and its appeal to specialized audiences, the large-scale field is defined by its proximity to the broader field of power and economic determinants.[4] Here, we might recognize the conventional opposition between 'high' and 'low' culture, and the symbolic positions occupied by avant-garde artists and commercial producers in the cultural field. [...]

In popular usage, the word 'glitch' has negative connotations. It refers to mechanical error or a rogue signal present within an electronic system and is conventionally seen as a problem. While its derivation from the Yiddish 'glitshn', to slip, slide or glide, suggests a physical movement, it is commonly used to describe errors in computer-based systems that result in a short electrical pulse. This meaning can be traced back to its usage by astronauts describing electrical malfunctions during the first US manned space-flight in 1962. Glitchy systems are systems prone to errors, the outcomes of which are often discernable as small audio spikes.

It is these sounds of error and related secondary audio phenomena such as static and interference that have become used as source material for musicians associated with the music style known as glitch. From the late 1980s, a cluster of bands such as Pan Sonic, Matmos and Oval, as well as a vast array of 'sound hackers' from Germany, Japan, the USA and elsewhere, turned to glitch as a way of creating and performing music. Drawing on the technological artefacts of error, as well as a rich history of avant-garde experimentation associated with *elektronische musik* and *musique concrète*, these musicians compose music as a series of micro incidents – bleeps, cuts, clicks and pulses – rendered by digital techniques and tools. [...]

It was during the 1990s that glitch really took off, however, as a steady increase in the amount of music produced under the category was matched by a visible expansion in the networks, discourses and accoutrements of glitch-related phenomena. Not only did glitch extend the sources of error to include computer-based system crashes, clipping and distortion, but the 'scene' had developed enough of a following and presence to warrant specialist CD compilations as well as a network of independent record labels based in Germany, France and the UK. Key support personnel such as critics gravitated to the category as the latest in a long line of (post)modern sonic interventions with counter-cultural connotations. Indeed, its intellectual appropriation as avant-digital deconstruction lent it the kind of leftfield gravitas so central to electronic arts festivals and specialist academic journals [...].

While glitch's origins in experimental art music are significant, however, its more recent dalliance with less restricted domains is also noteworthy. Just as electronica itself has become relatively normalized through channels of popular and consumer culture, so glitch has seeped into the mainstream via electronic and dance music festivals, film scores, radio airplay, as well as the odd car and mobile phone advert. Moreover, glitch bands such as Matmos, Autechre and Aphex Twin have attained a degree of popular (albeit far from superstar) appeal and coverage, and the style has very quickly crystallized into an established genre with recognizable gigs, stylistic signatures and labels – Mego, Touch, Thrill Jockey and Mille Plateaux to name just a few. [...]

This emphasis on the cerebral provocations of glitch is common in contemporary writings and directly maps onto the relatively small gap between musicians, critics and audiences. In most cases, glitch's support writers are themselves directly involved in the unfolding of the style, and their interventions are either internalist in content – fulfilling aesthetic, formalist or stylistic criteria – or posit glitch as somehow outside the field through the maintenance of a cool distance from pop. [T]he stylistic fundaments of glitch are [...] the latest in a series of socio-biological progressions dependent on imperfection.[5] [On] the other hand, glitch is a musico-philosophical intervention possessing a certain quality of alterity onto which are piled avant-garde aspirations towards cultural critique, shock and deconstruction.[6] [...] Powerful as these writings are, what is lost are the sets of social relations that make glitch-based interventions possible, as well as the broader socio-economic webs and networks that traverse them. Indeed, it is precisely because descriptive histories of glitch move us towards describing the style from within that we need to add the kinds of critical insights that Bourdieu's cultural sociology brings.

Prima facie, glitch fits Bourdieu's model rather well [...].

First, while it is stretched between the two poles of production [that is, within the overall Field of Cultural Production (Music), glitch stretches between the restricted sub-field 'Avant-garde/Experimental' on the one hand, and the large-scale sub-field 'Commercial Pop/Rock' on the other], glitch's aesthetic orientations are skewed towards the principles of the restricted sub-field. This is evident in glitch's connections with, and origins in, a rarefied world of academic computer music and audio research. In its 'pure' form, in fact, this kind of computer music stretches the limits of the possible forms of productive activity itself: that is, the limits of the field and therefore the limits of music, as it bleeds into 'sound', 'noise', 'research', and so on. Even in its less restricted forms, glitch music is aligned by critics and musicians to a kind of 'scientific approach' that yields experiments in form.[6] This is not to assume that it is somehow self-referential or 'outside', but that it owes its experimental dispositions to certain structural pre-conditions operative in the field itself.

Second, then, glitch's protagonists comprise a culturally-privileged fraction of specialists [with] high participation rates amongst graduates with humanities, music technology or sound design degrees, committed to innovation and autonomy. This commitment is made possible by the accumulated history of the field itself, including the positions occupied by previous electro-acoustic musicians – Pierre Schaeffer, Iannis Xenakis and John Cage being notable figures. Glitch audiences, on the other hand, tend to be drawn from a similarly restricted group of young educated technophiles and aficionados with a preference for experimental art music. [...]

Third, revered as the most recent example of sonic experimentation, glitch's aesthetic credentials are regularly defended with respect to both avant-garde practice and high theory, to the extent that critics and fans are as likely to evoke the work of Deleuze and Guattari and describe the genre as 'rhizomatic'.[7] as they are to place it in the pantheon of dance music. [...] Transposable inclinations between education and experimental music are revealed in a mastery of words and concepts around discourses of glitch. Here, the importance of the consecrating actions of influential critics is essential to the positioning of glitch as relationally distinct from mainstream pop. [...]

Like the dense explanations accompanying contemporary art, writings on glitch are essential to its symbolic legitimacy. Liner notes, in particular, are a key site of its discursive presence, with a typical compilation bearing a dense accompaniment of essays packed with Deleuzian allusions. Notwithstanding the question of whether anyone reads these notes, their mere presence is testament to a belief among those equipped with the esoteric codes that glitch is distinct. This is why criticism is the site of an 'objective connivance'.[8] between readers and critics, as the homology between the two is also a structural correspondence between the intellectual field and the readership's location within a dominant class field. [...]

[W]hile still part of an ever fragmenting music industry, glitch is closer to what Bourdieu calls a 'research sector',[9] its position secured by an opposition to the incumbents of more commercial styles of music dubbed as 'easy' or 'formulaic'. Even the physical sites of some glitch performances are telling, many becoming the staple diet of classical concert halls and avant-garde galleries. A recent performance by Ryoji Ikeda at the Sage at Gateshead (in northern England), for instance, had all the trappings of a classical sojourn for the refined cultural intelligentsia, the purpose-built concert hall graced by an audience that would not have looked out of place at a Schoenberg concert.

Boundaries between sub-fields are not impermeable, however. Indeed, as Bourdieu himself states: 'one must be wary of establishing a clear boundary, since they are merely two poles, defined in and by their antagonistic relationship, of the same space'.[10] Fourth, then, an interesting recent development has been the way glitch as a technique and style has migrated into more commercial forms of music. Both Björk and Radiohead have appropriated glitchy sounds for their own works with some degree of commercial success. In the case of her 2001 album Vespertine, Björk even went so far as to call in the specialist glitch band Matmos for programming duties on three of her tracks, while Madonna's hit 'Don't Tell Me' (2000) contains an array of glitchy interruptions. Unsurprisingly, mainstream artists and producers have picked up on the fact that glitch carries with it bleeding-edge connotations. Just as 'cool', 'edge' and 'risk' have become commodified offshoots of the domestication of the avant-garde, so glitch is becoming one of the latest targets in a long succession of outré styles considered fair game for appropriation. Indeed, a host of software companies are already coding glitch-making 'plug-ins' that automatically produce the sounds of computer error to order, without the musician having to slice their CDs or tinker with the insides of computers. This accommodation and commercialization reprises the historical trajectory of vinyl scratch (itself once considered radical and annoying) from the likes of Grandmaster Flash and Christian Marclay into the pop mainstream. [...]

We can begin to see, then, how the field concept can 'sociologize' spheres of cultural practice in important ways. As an overall map of the terrain of culture and its dialogues with power, the field orients us to positional co-ordinates and their logics. It shows us how alliances and differentiations really matter in the making of movements, genres and styles [...]. It finesses our approach to the music world by describing how the cultural field is internally configured according to a series of associations and schisms between genres, institutions and associated personnel. It also makes good sociological sense of the preconditions of autonomy, moving us away from statements that affirm the cultural 'independence' of musicians without reflecting on how this independence is actually a profound dependence on the joint histories of habitus and field. As for broader, contextual issues, the field concept gives us analytical purchase on the mechanism by which spheres of practice like music re-translate the incursions of economic and political forces. [...]

And yet for all these benefits, there is still something missing. No concept, of course, is perfect, but cracks are beginning to show in Bourdieu's ideas.[...] Bourdieu rarely addresses technology. [...]

Well, perhaps [there] are just missing details and oversights. After all, Bourdieu cannot be expected to cover everything. Indeed, it is entirely possible to construct a Bourdieu-inspired take on technology without too much difficulty. [...] In the case of glitch, for instance, we could quite easily plot how hierarchies of capital correspond to different choices and uses made of particular technologies such as software applications. This would follow a logic whereby the more popular an application is, the less likely it will be adopted in good faith by avant-gardistes. We could also examine how techniques of authoring highlight dispositional consumption practices and aesthetic investments in the field: does the musician use samples from mass-produced CDs or generate their own audio material in the field? Do they use preset sounds bundled with software applications or programme their own sounds and patches? Do they use a Mac – the sine qua non of the 'culturally enlightened' – or a PC?

These are all lines of inquiry befitting a field approach. Still, if we keep asking these questions, we are left continuously rehearsing the mantra of technology as an instrumental 'badge' or a 'thing' that secures and reproduces. Technologies rarely open up, they close down; they are technical and symbolic resources, extra weapons in the game. What is missing here, I would suggest, is the texture of technology, not just in relation to the more phenomenological aspects of tweaking and twiddling but also to the multifarious modifications and translations that technologies afford, to their efficacy beyond reproduction, to what they make possible. [...]

It has been an insight of actor network theorists (ANTs) such as Bruno Latour and Michel Callon to explain the importance of non-human actors in the social world. Actors such as chemicals, airbags and door knobs impose their presence in all sorts of ways that make them partners in interaction. This means that action is no longer perceived as the sole realm of the human actor, but also the realm of the non-human actor, including the technological artefact. For their presence, the world is not exactly as it was before, a positivity has been made that changes the course of events. This position adds to Howard Becker's notion of an 'art world'[11] as collective activity the important understanding that techniques, settings and devices exchange their properties with humans. These entities comprise a cluster of elements – inanimate and animate – that might at any point add their identities and relationships into the collective.[12]

When one opens action up like this, the points of articulation and influence between a range of entities are enlarged such that 'production' becomes a full and expansive concept. It also allows for a degree of slippage between the prescriptions encoded in the manufacture of artefacts [...] and the unforeseen uses that these technologies end up affording through breakdown, error and misuse. Indeed, the history of music bulges with cases that point to the unpredictable, productive and unstable: turntables as DJ instruments, monophonic bassline generators such as the Roland TB303 mis-programmed to beget acid house, telephone bandwidth-saving technologies turned into vocoders.

Perhaps the computer itself might be a case where digital audio flexibility and the increasing availability of music software sends all sorts of forces into the practices of music making and the inevitability of new forms and genres. After all, glitch is glitch (and not grunge, hip hop, trip hop or drum and bass) not just because of its field position as conventionally understood by Bourdieu; not just because of the habitus-derived uses its protagonists have made of hardware and software; but also because of these technologies themselves. That is to say, the gathering of digital objects around glitch changes not only how the music is made, but also what the music 'is'. The codes, the coding, the graphic user interface, the CDs, the various hardware interfaces and their design – these all make a difference. They do not determine the style alone, but neither are they merely a backdrop to, or weapon for, the purposeful action of the acquisitive human actor. They are objects essential to the relay of social relations in the formation of glitch to the same extent as non-human objects are in the formation of all styles and genres. We might, indeed, speculate that Bourdieu fails to tackle non-human objects head on precisely because they introduce elements of presence, uncertainty and deviation into fields in a way that poses a series of problems to Bourdieu's own schema .

One does violence to the intricacies of the social world when technology is framed as a passive recipient, tool or 'subset' of the habitus enacted in fields,[13] (and not also an active force in those fields. [...] This recognition is important in a contemporary context where musicians enter into increasingly immersive relations with their instruments and form increasingly complex machine-body assemblages. Yet, throughout music history, as Antoine Hennion has shown, material devices such as scores, concert staging, acoustic treatments and musical texts have always formed an 'interconnected series of mediations ... creating an irreversible movement which none of them alone would have been able to achieve'.[14] [...]

One needs to be guarded against work that claims a self-organized, machinic evolution and genesis of technology independent of its uses and meanings amidst social spheres of practice. This is precisely the reason why the best work in actor network theory alerts us to how the technical and the social are inextricably linked, in turn sensitizing us to the fact that instruments and associated devices are not passive intermediaries but active mediators. [...]

To return to the case of glitch, then, one needs to hold together objects, trajectories and materials without losing sight of its socially organized formation in successive phases of attachment and opposition. The field clearly does set certain limits, particularly in how specific modes of operation and intervention among glitch musicians are played out, but glitch is also held together by an array of other objects which populate these relations and without which the style becomes unthinkable: transistors, electrical pulses, keyboards, software, graphic user interfaces, laptops, CDs, digital signal processing tools, the internet. It is the latter, for instance, that has been the proliferating condition of knowledge under which glitch musicians have learned the tricks of the trade, including how to use and abuse particular forms of software; it is an earlier breakdown in the material properties of technologies such as the CD and CD player that gave glitch its source materials; and it is the visual streams and blocks of MIDI data in applications such as Ableton Live, Max/MSP and AudioMulch that have enjoined the glitch musician in new modes of working.

[...] Indeed, what better way of making sense of the way glitch represents an intermeshing of humans and technologies at one and the same time as it has accreted symbolic validity in a field of relations than to strategically deploy insights from both Bourdieu and ANTs. [...S]uch a commitment to glitch's materiality is not to ignore the position-takings of the musicians, critics and labels themselves, but to examine human and non-human materials as co-producers of the field, as heterogeneous assemblages ongoingly exchanging their properties in relatively structured settings: to open the black box of technology as well as the well-regulated ballet of the field.

1 Pierre Bourdieu, *The Rules of Art* (1992) (Cambridge: Polity Press, 1996).

2 Pierre Bourdieu, *In Other Words: Essays towards a Reflexive Sociology* (1987) (Cambridge: Polity Press, 1990) 87.

3 Pierre Bourdieu and Loïc Wacquant, *An Invitation to Reflexive Sociology* (Cambridge: Polity Press, 1992).

4 Ibid., 145.

5 Torben Sangild, 'Glitch: The Beauty of Malfunction', in C.J. Washburne and M. Derno, eds, *Bad Music* (London and New York: Routledge, 2004) 257–74.

6 [footnote 7 in source] Rob Young, 'Transformed by Sound: Autechre', *The Wire*, no. 156 (May 1997).

7 [8] Janne Vanhanen, 'Loving the Ghost in the Machine', *Ctheory* (26 November 2001) 8. (www.ctheory.net)

8 [9] Pierre Bourdieu, *The Field of Cultural Production* (Cambridge: Polity Press, 1993) 94.

9 [10] Pierre Bourdieu, *The Rules of Art*, op. cit., 120,

10 [11] Ibid., 120.

11 [12] Howard Becker, *Art Worlds* (Berkeley and Los Angeles: University of California Press, 1982).

12 [13] Michel Callon, 'Society in the Making: The Study of Technology as a Tool for Sociological Analysis', in W.E. Bijker, et al., eds, *The Social Construction of Technological Systems* (Cambridge, Massachusetts: The MIT Press, 1987) 83–103.

13 [14] Jonathan Sterne, 'Bourdieu, Technique and Technology', op. cit., 370.

14 [15] Antoine Hennion, 'Baroque and Rock Music, Mediators and Musical Taste', *Poetics*, vol. 24, no. 6 (2007) 424.

Nick Prior, extracts from 'Putting a Glitch in the Field: Bourdieu, Actor Network Theory and Contemporary Music', *Cultural Sociology*, vol. 2, no. 3 (November 2008) 303–17.

Niklas Luhmann
The Function of Art and the Differentiation of the Art System//1995

[...] Today, systems theory is a highly developed, albeit controversial, analytical instrument. It requires theoretical decisions that do not directly concern art. (This, of course, holds for other – for example, semiological – analyses of art as well.) In conjunction with the thesis that society is a functionally differentiated system and is in this form historically unique, a systems-theoretical orientation has further consequences. It means that the different functional systems are treated in many respects as comparable. [...] Issues such as system formation and system boundaries, function, medium and forms, operative closure, autopoiesis, first and second-order observation, and coding and programming can be investigated with regard to any functional system. As these investigations take shape and yield answers, a theory of society emerges that does not depend on discovering a unified meaning behind society – for example, by deriving societies from the nature of man, from a founding contract, or from an ultimate moral consensus. Such propositions may be treated as part of the theory's subject matter, as different forms of self-description available to the system of society. What ultimately characterizes society, however, manifests itself in the comparability of its subsystems. [...]

In a domain such as art (just as for law, science, politics, and so on), we discover not unique traits of art but features that can be found, *mutatis mutandis*, in other functional systems as well – for example, the shift to a mode of second-order observation. Art participates in society by differentiating itself as a system, which subjects art to a logic of operative closure – just like any other functional system. [...] Modern art is autonomous in an operative sense. No one else does what it does. The societal nature of modern art consists in its operative closure and autonomy. [...]

We base the following analyses on a distinction, namely, on the distinction between system/environment relations, on the one hand, and system/system relations, on the other. When dealing with system/environment relations, the system constitutes the internal side of the form, whereas the environment is its unmarked space. [...] If, however, we are dealing with system/system relations, then the other side can be marked and indicated. In this case, art no longer deals with 'everything else' but with questions such as whether and to what extent the artist is motivated by political convenience or by wealthy customers. [...]

What happens to art if other social domains, such as the economy, politics, or science, establish themselves as functional systems? What happens when they focus more narrowly on a special problem, begin to see everything from this perspective, and eventually close themselves off with an eye toward this problem? What is art if in fourteenth-century Florence the Medicis support art as a way of politically legitimizing money acquired in dubious ways, which they subsequently invest in consolidating their political position? What happens to art if the functionally oriented differentiation of other systems pushes society as a whole toward functional differentiation? Will art become the slave of other functional systems, which dominate from now on? Or does – as indeed we shall argue – the increasing automatization of functional systems challenge art to discover its own function and to focus exclusively on this function? [...]

[W]e must formulate more radically the difference that art establishes in the world. [...] One might start from the assumption that art uses perceptions and, by doing so, seizes consciousness at the level of its own externalizing activity. The function of art would then consist in integrating what is in principle incommunicable – namely, perception – into the communication network of society. [...] Kant already located the function of art (of the presentation of aesthetic ideas) in its capacity to stimulate thinking in ways that exceed verbal or conceptual comprehension. The art system concedes to the perceiving consciousness its own unique adventure in observing artworks – and yet it makes available as communication the formal selection that triggered the adventure. [...]

An independent relation between redundancy and variety characterizes perception. In a manner that is matched neither by thought nor by communication, perception presents *astonishment and recognition* in *a single instant*. Art uses, enhances, and in a sense exploits the possibilities of perception in such a way that it can present the *unity of this distinction*. To put it differently, art permits observation to oscillate between astonishment and recognition [...] for example, by quotations from other works that render repetitions at once familiar and strange [...] However [...] the identification of repetition relies on perception rather than on conceptual abstraction. Art specializes in this problem, and this distinguishes it from ordinary efforts to cope with small irritations in everyday perception. [...]

The work of art, then, establishes a reality of its own that differs from ordinary reality. And yet, despite the work's perceptibility, despite its undeniable reality, it simultaneously constitutes another reality, the meaning of which is imaginary or fictional. Art splits the world into a real world and an imaginary world in a manner that resembles, and yet differs from, the use of symbols in language or from the religious treatment of sacred objects and events. The function of art concerns the meaning of this split – it is not just a matter of enriching a given world with further objects (even if they are 'beautiful'). [...]

The imaginary world of art offers a position from which *something else* can be determined *as reality* – as do the world of language, with its potential for misuse, or the world of religion, albeit in different ways. Without such markings of difference, the world would simply be the way it is. Only when a reality 'out there' is distinguished from fictional reality can one observe one side from the perspective of the other. Language and religion both accomplish such a doubling, which allows us to indicate the given world *as real*. Art adds a new twist to this detour, which leads via the imagination away from and back to reality – art realizes itself in the realm of perceptible objects. Any other doubling of reality can be copied into the imaginary reality of the world of art – the doubling of reality and dream, for example, of reality and play, of reality and illusion, even of reality and art. Unlike language and religion, art is made, which implies freedoms and limitations in the choice of forms unknown to language and religion. [...]

Only within a differentiated distinction between a real and a fictional, imagined reality can a specific relationship to reality emerge, for which art seeks different forms – whether to 'imitate' what reality does not show (its essential forms, its ideas, its divine perfection), to 'criticize' reality for what it does not want to admit (its shortcomings, its 'class rule', its commercial orientation), or to affirm reality by showing that its representation succeeds, in fact, succeeds so well that creating the work of art and looking at it is a delight. The concepts imitation/critique/affirmation do not exhaust the possibilities. Another intent might address the observer as an individual and contrive a situation in which he faces reality (and ultimately himself) and learns how to observe it in ways he could never learn in real life. [...]

The question might be rephrased as follows: How does reality appear when there is art?

In creating a double of reality from which reality can be observed, the artwork can leave it to the observer to overcome this split – whether in an idealizing, critical or affirmative manner, or by discovering experiences of his own. Some texts are meant to be affirmative and oppose the hypercritical addiction to negativity – yet they can be read in an ironical or melancholy mode, or as mirroring one's own experiences with communication. [...] Because it embeds its forms in objects, art need not enforce a choice between consensus and dissent, or between an affirmative and a critical attitude toward reality. Art needs no reasonable justification, and by unfolding its power of conviction in the realm of perceptible objects, it demonstrates this. The 'pleasure' afforded by the artwork, according to traditional doctrine, always also contains a hint of malicious joy, indeed of scorn, directed against the vanity of seeking access to the world through reason. [...]

An independently developed sense of form in art leads to gains in autonomy, especially when art develops its own dynamics and begins to react to itself. [...] Everyday life becomes worthy of art, and what used to be significant is subjected to distorting misrepresentations. [...] Common values were not just negated or turned on their head; they were neutralized and rejected as distinctions for the sake of demonstrating possibilities of order that had nothing to do with them. [...] Against these trends, art developed procedures and principles of its own – novelty, obscurity, style-consciousness, and eventually a self-description that thematizes the various artistic genres and sets them apart from the new rationalism. [...]

In the twentieth century, one encounters artworks that seek to cancel the difference between a real and an imagined reality by presenting themselves in ways that make them indistinguishable from real objects. [...] No ordinary object insists on being taken for an ordinary thing, but a work that does so betrays itself by this very effort. The function of art in such a case is to reproduce the difference of art. But the mere fact that art seeks to cancel this difference and fails in its effort to do so perhaps says more about art than could any excuse or critique. Here, what we learn to observe is the inevitable and ineradicable rule of difference. [...]

Unlike philosophy, art does not search for islands of security from which other experiences can be expelled as fantastic or imaginary, or rejected as a world of secondary qualities or enjoyment, of pleasure or common sense. Art radicalizes the difference between the real and the merely possible in order to show through works of its own that even in the realm of possibility there is order after all. [...]

Within the gravitational field of its function, modern art tends to experiment with formal means. The word *formal* here does not refer to the distinction, which at first guided modern art, between form and matter or form and content, but to the characteristics of an indicating operation that observes, as if from the corner of its eye, what happens on the other side of form. In this way, the work of art points the observer toward an observation of form. This may have been what was meant by the notion 'autotelic'. However, the social function of art exceeds the mere reconstruction of observational possibilities that are potentially present in the work. Rather, it consists in *demonstrating the compelling forces of order in the realm of the possible*. [...]

Art raises the question of whether a trend toward 'morphogenesis' might be implied in any operational sequence, and whether an observer can observe at all except with reference to an order – especially when observing observations.

From this perspective, the *formal complexity* a work is *capable of achieving* becomes a crucial, indeed, the decisive variable. Whatever functions as the other side of a form requires decisions about further forms that generate other sides of their own, which raises the problem of how much variety the work's recursive integrity can accommodate and keep under control. [...] Contrary to widely held notions, the function of art is not (or no longer) to represent or idealize the world, nor does it consist in a 'critique' of society. Once art becomes autonomous, the emphasis shifts from hetero-reference to self-reference – which is not the same as self-isolation, not *l'art pour l'art*. Transitional formulations of this type are understandable. But there is no such thing as self-reference (form) without hetero-reference. And when art displays a self-positing order in the medium of perception or imagination, it calls attention to a logic of reality which expresses itself not only through the real but also in fictional reality. Within the difference real/fictional reality, the unity of the world (the unity of this difference) escapes observation by presenting itself as the order of the distinction's form.

Art has no ambition to redeem society by exercising aesthetic control over an ever-expanding realm of possibility. Art is merely one of society's functional systems, and even though it may harbour universalistic ambitions, it cannot seriously wish to replace all the other systems or force these systems under its authority. The functional primacy of art holds exclusively for art. This is why, protected by its operative closure, art can focus on its own function and observe, from within ever-expanding boundaries, the realm of possibility with an eye toward fitting form combinations. [...]

When Hegel speaks of the end of art [...] he can mean only one thing: art has lost its immediate relation to society and worldly affairs and must henceforth acknowledge its own differentiation. Art can still claim universal competence for almost everything, but it can do so only *as art* and only on the basis of a specific mode of operation that follows its own criteria.

The notion that art, as represented by artists, can find a knowledgeable and sympathetic counterpart somewhere else in society must be sacrificed as well. A supporting context – if this is what one is looking for – is no longer available. A model based on complementary roles for artists and connoisseurs can no longer represent the couplings between the art system and society. Rather, it represents the differentiation of art as communication in society. The interaction between artists, experts and consumers differentiates itself as communication, and it takes place only in the art system, which establishes and reproduces itself in this manner. What romanticism called 'art criticism' is integrated into the art system as a 'medium of reflection', and its task is to complete the artist's work. [...] What it actually reflected upon, however, is the autonomy imposed upon art – the functional differentiation of society. [...]

The differentiation of the art system – a process characterized simultaneously by continuity and discontinuity – allows the relation between system and environment to be reintroduced into the system in the form of a relationship between self-reference and hetero-reference. As we recall, there can be no self-reference without hetero-reference, for it is not clear how the self can be indicated if it excludes nothing. When the unity of self-reference and hetero-reference becomes an issue, searching for the common denominator in the meaning of reference suggests itself: what is the reference of 'reference'? [...]

Depending on how the relationship between self-reference and hetero-reference is applied, we shall distinguish an art that is primarily *symbolic* from an art that thinks of itself as a *sign*, and we shall further distinguish an art that specializes in experimenting with *form combinations*. Prior to its differentiation, art was considered *symbolic* if it searched for a higher meaning in its condensed ornamental relationships. In the course of the court- and market-oriented phases of its differentiation, art turned into a *sign*. The sign, by virtue of what was believed to be its objective reference, stood for what the artist, the connoisseur and the lover of art had in common. But once the differentiation of this community was realized as communication, the only remaining option was to observe the continual balancing between self-reference and hetero-reference in the operations of the art system. Under these conditions, one finds the nexus between self- and hetero-reference in the *formal combinations* of artworks that facilitate an observation of observations. [...]

Niklas Luhmann, extracts from *Art as a Social System* (Frankfurt am Main, 1995); trans. Eva M. Knodt (Stanford: Stanford University Press, 2000) 133–83 [footnotes not included].

Christian Katti
Systematically Observing Surveillance: Paradoxes of Observation According to Niklas Luhmann's Systems Theory//1999

Whispered voice-over in parentheses: (Look into the camera ... we see you.) (Who's there?) ... He feels in control (camera) for the first time (he is the camera). The camera cuts to a floating, circling shot above the bed. He lies there on the covers (early American); a fluorescent pillow rings his head (halo). His young frame folds together now into a foetal position ...
– Tony Oursler, *White Trash* (1993)

[...] The paradox inherent to surveillance phenomena is well known. The attractiveness of problems of paradox, however, earns a very positive estimation in Niklas Luhmann's perspective, whereas repressing and avoiding paradoxes does not really help us out of the pitfalls with which they are connected. Paradoxes emerge [...] when the conditions of possibility of an operation are also the conditions of its impossibility. Paradoxes are problems par excellence for every observer, but not necessarily for the operations of observation, which instead experience a dynamization by means of paradoxes. This dynamic makes paradoxes apparently productive, but what do we gain from them besides the confusion we know all too well? 'The conventional answer seems to be: exercise of wit.'[1] And how can wit and understanding be sharpened and trained? The exercise is 'to deframe and reframe the frame of normal thinking, the frame of common sense.'[2] The communication of paradoxes fixes attention on the frames of common sense, frames that normally go unattended. If this is the function then it will not surprise us that deframing again needs its own frames.' Art appears to be predestined [...] for this task of multifaceted framing and destabilization. Indeed, art is capable of enabling experiments with reality and even with the way in which we perceive this reality. With a reflexive turn, it allows for perceptions about *how* we perceive reality. Seen ambitiously, as is usual in the tradition of modernity, art should still be conceived of autonomously, which frees it from tasks imposed from the outside, no matter how useful or sublime. It is one of the strengths of Luhmann's theory of art that it conceives of art as autonomous and at the same time as a socially and historically constituted phenomenon. All sub-systems of society are autonomous, according to Luhmann; otherwise they would not be able to differentiate themselves as sub-systems. [...] The characteristics of observation that I will outline are relevant to areas of (as well as beyond) art. They have this in common with the phenomena of surveillance, which can be located in aesthetic contexts as well as in other areas, taking on various roles according to the situation.

Besides the strict separation of psychical and social systems, and the result that society does not consist of psychical systems but rather of communication, Luhmann's theory has us believe that observations are not directly communicable and communication, no matter what kind, is not directly observable. Of course, [the] specialized notions of 'communication' and 'observation' used here warrant further explanation [...]. Indeed, this point is already a demonstration of the ideological potential – one almost wants to say that it is determined by Zeitgeist and culture – of surveillance phenomena and methods, especially when they are working with technological means like photography, video and audio recording. The technological medium produces something that is generally accepted as a means of providing evidence. In this mediatized condition, observations are apparently directly translatable and exchangeable – but only apparently. The more the inner connective energies of societies are conceived of in dissolution, the more emphatic the power of proof and the violence of (producing) evidence of these apparently exchangeable observations are becoming. Yet we do not want to get ahead of ourselves here, since Luhmann's concept of 'medium' is linked to a concept of 'form', and it is not subsumed in the generally accepted definition of 'medium', which is usually only considered in the technological sense. [...]

The general assumption of this essay is that every kind of surveillance is necessarily linked with observation. Put negatively: no surveillance can occur without the execution of operations of observation. This somewhat unspectacular assertion [makes] clear that every act of surveillance necessarily produces its opposite. Surveillance and observation result in something that one can call a 'blind spot' analogous to that of the eye; the blind spot is something that surveillance and observation cannot see, cannot observe, for systematic reasons. [...] The indissoluble double gesture of revealing and concealing, veiling and apparently unveiling, has a paradoxical quality [...]. In order for something to be made observable at all, other things – certain fields that are ambiguously linked to observation and organize it in a certain way – drop out of the same observation. In brief, the paradox emerges that by means of producing something (an observation), we unwillingly also produce its opposite (concealing). What results from this for the plethora of surveillance phenomena and their effects is a not insignificant question. The fact that totalitarian fantasies of surveillance, dreaming of complete transparency, have also to observe themselves, thereby rendering themselves opaque, is the irony and paradox of the field outlined here. And it is hardly sufficient to downplay *Real*-political phenomena of this kind as an 'irony of history'.

When a second observer sees what the first one does not see – thus, when he observes an observer observing – this is called second-order observation. This observation of observation sees 'what the observer sees and how he sees what he sees. It even sees what the observed observer does not see, and sees that he does not see what he does not see.'[3] But in order to see all of that, the second-order observation has to use a distinction that remains invisible to it, since only by means of this distinction is it able to observe what it observes. In terms of this operatively applied distinction, a second-order observation is also a first-order observation, which also results in a paradox, since it is simultaneously something and something else. 'Each [observer] observes what he is capable of on the basis of his own paradox, invisible to himself, and based on a distinction whose unity is inaccessible to his observation.'[4]

Here observation is conceived not in terms of the human being but rather systemically, 'highly abstractly and independently of the material substrate, the infrastructure or the specific manner of observation.'[5] [...]

An observer can observe it/himself observing, just not simultaneously; the observer must use an operation of higher-order observation, and then the latter is unobservable. [...] The observer observes himself as another.

This point is also foregrounded by Peter Weibel's closed-circuit video installation *Beobachtung der Beobachtung: Unbestimmtheit* (Observing Observation: Uncertainty) (1973), in which the observer, located in the inner space of observation, 'can never see [him or herself] from the front, no matter how much [the observer] twists and turns'.[6]

The three cameras aimed at him/her and the monitors connected to them systematically prevent the viewer from seeing a front view of himself/herself in the monitor. This calculated effect of the view of oneself from behind is constructed by the arrangement of the cameras and monitors, and it becomes especially apparent when the observer moves within the installation in order to escape being trapped in this rear view, which is, however, not possible. And all of this may become the 'subject' for observers and observations of second order and higher order that still are part of the installation, even more so when they are not located/taking place in the inner sphere. [...]

By means of second-order cybernetics, Luhmann is able to explain that 'operations of "subjects" often are best understood when one considers them to be induced by observation, thus, brought about when the observed object itself is functioning as an observer. The distinction subject/object is thus implemented neither naturally nor transcendental-theoretically via self-reflection of the consciousness, but rather it is a distinction that is testing itself in the praxis of observation; it can be applied not only to humans but also to animals and social systems, and perhaps even to electronic machines, *whenever the complicated two-part operation of observing observers succeeds.*'[7] [...]

1 [footnote 3 in source] Niklas Luhmann, 'The Paradoxy of Observing Systems', *Cultural Critique* (Fall 1995) 39.

2 [4] Ibid.

3 [15] Niklas Luhmann, 'Identität- was oder wie?', *Soziologische Aufklarung 5: Konstruktivistische Perspectiven* (Opladen: Westdeutscher Verlag, 1990) 16.

4 [16] Niklas Luhmann, 'Sthenography', trans. Bernd Widdig, *Stanford Literature Review*, vol. 7, no. 1–2 (Spring–Fall, 1990) 137.

5 [17] Niklas Luhmann, *Die Gesellschaft der Gesellschaft*, 2nd ed. (Frankfurt am Main: Suhrkamp, 1999) vol 1., 69.

6 [20] Peter Weibel, *Mediendichtung, Protokolle, 2/1982* (Vienna and Munich: Jugend und Volk, 1982) 118.

7 [43] Niklas Luhmann, 'Ich sehe was was Du nicht siehst', *Soziologische Aufklärung 5*, op cit., 232. Luhmann's italics.

Christian Katti, extracts from 'Systematically Observing Surveillance: Paradoxes of Observation According to Niklas Luhmann's Systems Theory' (1999), in *CTRL [SPACE]: Rhetorics of Surveillance from Bentham to Big Brother*, ed. Thomas Y. Levin, et al. (Karlsruhe: ZKM/Cambridge, Massachusetts: The MIT Press, 2002) 51–63.

Bruno Latour
Some Experiments in Art and Politics//2011

The word 'network' has become a ubiquitous designation for technical infrastructures, social relations, geopolitics, mafias and, of course, our new life online. But networks, in the way they are usually drawn, have the great visual defect of being 'anaemic' and 'anorexic', in the words of philosopher Peter Sloterdijk, who has devised a philosophy of *spheres* and *envelopes*. Unlike networks, spheres are not anaemic, not just points and links, but complex ecosystems in which forms of life define their 'immunity' by devising protective walls and inventing elaborate systems of air conditioning. Inside those artificial spheres of existence, through a process Sloterdijk calls 'anthropotechnics', humans are born and raised. The two concepts of networks and spheres are clearly in contradistinction to one another: while networks are good at describing long-distance and unexpected connections starting from local points, spheres are useful for describing local, fragile and complex 'atmospheric conditions' – another of Sloterdijk's terms. Networks are good at stressing edges and movements; spheres at highlighting envelopes and wombs.

Of course, both notions are indispensable for registering the originality of what is called 'globalization', an empty term that is unable to define from which localities, and through which connections, the 'global' is assumed to act. Most people who enjoy speaking of the 'global world' live in narrow, provincial confines with few connections to other equally provincial abodes in far away places. Academia is one case. So is Wall Street. One thing is certain: the globalized world has no 'globe' inside which it could reside. As for Gaia, the goddess of the Earth, we seem to have great difficulty housing her inside our global view, and even more difficulty housing ourselves inside her complex cybernetic feedbacks. It is the globe that is most absent in the era of globalization. Bad luck: when we had a globe during the classical age of discoveries and empire, there was no globalization; and now that we have to absorb truly global problems ...

Tomás Saraceno's *Galaxies Forming along Filaments*

So how can we have both networks and spheres? How do we avoid the pitfalls of a globalization that has no real globe in which to place everything? In a work presented at the Venice Biennale in 2009, Tomás Saraceno provided a great, and no doubt unintended, metaphor for social theory. In an entire room inside the Biennale's main pavilion, *Galaxies Forming along Filaments, Like Droplets along the Strands of a Spider's Web* (2008) consisted of carefully mounted elastic connectors that produced the shape of networks and spheres. If you were to avoid the guards' attentive gaze and slightly shake the elastic connectors – strictly forbidden – your action would reverberate quickly through the links and points of the network paths, but much more slowly through the spheres. This is not to say that spheres are made from different stuff, as if we must choose between habitation and connection, between local and global, or indeed between Sloterdijk and, let's say, actor-network theory. What Saraceno's work of art and engineering reveals is that multiplying the connections and assembling them closely enough will shift slowly from a network (which you can see through) to a sphere (difficult to see through). Beautifully simple and terribly efficient. [...]

Saraceno performed precisely the task of philosophy according to Sloterdijk, namely of *explicating* the material and artificial conditions for existence. The task is not to overthrow but to make explicit. [...] *Galaxies Forming along Filaments* allows those who try to redescribe the loose expression of globalization to explore new concepts. Instead of having to choose between networks and spheres, we can have our cake and eat it too. There is a principle of connection – a kind of movement overlooked by the concepts of networks and spheres alike – that is able to generate, in the hands of a clever artist, both networks and spheres; a certain topology of knots that may thread the two types of connectors in a seamless web.

More interesting still is the theory of envelopes – the concept implied by this percept. In this proposition, walls or quasi-walls are supported by both external and lateral linkages. Again, we all know, or should know, that identities – the walls – are made possible only through the double movement of connecting distant anchors and stitching together local nodes. If you believe that there are independent bubbles and spheres that can sustain themselves, you are clearly forgetting the whole technology of envelopes. But it is one thing to say it, for instance in political philosophy – that no identity exists without relations with the rest of the world – and it is quite another to be reminded *visually* and *experientially* of the way this could be done.

Standing in the middle of Saraceno's work, the experience is inescapable: the very possibility of having an envelope around a local habitat is given by the length, number and solidity of the connectors that radiate out in all directions. I would have loved to see, when the exhibition was dismantled, how quickly the spherical patterns would have collapsed once a few of their outside links had been severed. A powerful lesson for ecology as well as for politics: the search for identity 'inside' is directly linked to the quality of the 'outside' connection – a useful reminder at a time when so many groups clamour for a solid identity that would 'resist globalization', as they say. As if being local and having an identity could possibly be severed from alterity and connection.

Another remarkable feature of Saraceno's work is that such a visual experience is not situated in any fixed ontological domain, nor at any given scale: you can take it, as I do, as a model for social theory, but you could just as well see it as a biological interpretation of the threads that hold the walls and components of a cell, or, more literally, as the weaving of some monstrously big spider, or the utopian projection of galactic cities in 3-D virtual space. This is very important if you consider that all sorts of disciplines are now trying to cross the old boundary that has, until now, distinguished the common destiny of increasing numbers of humans and non-humans. No visual representation of humans as such, separated from the rest of their support systems, makes any sense today. This was the primary motive for Sloterdijk's notion of spheres, as well as for the development of actor-network theory; in both cases the idea was simultaneously to modify the scale and the range of phenomena to be represented, so as to renew what was so badly packaged in the old nature/society divide. If we have to be connected with climate, bacteria, atoms and DNA, it would be great to learn about how those connections could be represented.

The other remarkable feature of the work is that although there are many local orderings – including spheres within spheres – there is no attempt at nesting all relations within one hierarchical order. There are many local hierarchies, but they are linked into what appears visually as a heterarchy. Local nesting, yes; global hierarchy, no. For me, this is a potent attempt at shaping today's political ecology – by extending former natural forces to address the human political problem of forming livable communities. [...]

To think in these terms is to find a way to avoid modernism – in which case the hierarchy moves from bigger to smaller elements from a central point – but also to avoid, if I dare say, postmodernism – in which case there would be no local hierarchies and no *homogeneous* principle by which to establish the connections (in this case the elastic tensors that provide the language for the whole piece). For me, that is the beauty of Saraceno's work: it gives a sense of order, legibility, precision and elegant engineering, and yet has no hierarchical structure. It is as if there were a vague possibility of retaining modernism's feeling of clarity and order, but freed from its ancient connection with hierarchy and verticality.

Who Owns Space and Time?

To explore the artistic, philosophical and political questions raised by Saraceno's work, it might be useful to turn to another *locus classicus* – not the sphere versus network debate, but the debate over who owns the space in which we live collectively. There is no better way to frame this question than the bungled dialogue [...] between Henri Bergson and Albert Einstein in Paris in 1922. After Bergson spoke for thirty minutes, Einstein made a terse two-minute remark, ending with this damning sentence: 'Hence there is no philosopher's time; there is only a psychological time different from the time of the physicist.' While Bergson had argued that his notion of space and time had a *cosmological* import that was to be carefully meshed within Einstein's remarkable discoveries, Einstein argued that there was only one time and space – that of physics – and that what Bergson was after was nothing more than *subjective* time – that of psychology. We recognize here the classical way for scientists to deal with philosophy, politics and art: 'What you say might be nice and interesting but it has no cosmological relevance because it only deals with the subjective elements, the lived world, not the real world.' [...]

Can we do better at the beginning of the twenty-first century? In other words, is it possible to give Bergson another chance to make his case that, no, he is not talking about subjective time and space, but is rather proposing an alternative to Einstein's cosmology? To explore such a possibility, I decided to rely on the fascinating genre of the reenactment. As many artists have shown, especially Rod Dickinson in the amazing staging of Milgram's experiment, reenactment is not a mere facsimile of the original but a second version, or a *second print* of the first instance, allowing for the exploration of its originality. This is why, in a series of lectures at the Centre Pompidou in June 2010, I invited, among many others, the artist Olafur Eliasson and two scholars, a historian of science, Jimena Canales, and a philosopher, Elie During, to reenact the famous debate by allowing the conclusion to shift somewhat, thus reopening a possibility that had been closed in the twentieth century.

Who *owns* the concepts of space and time? Artists? Philosophers? Scientists? Do we live in the space-time of Einstein without realizing it, or, as Bergson vainly argued, does Einstein, the physicist, live in the time of what Bergson called *duration*? Those questions, it seemed to me, were just as important for physicists, historians, and philosophers as they are for an artist like Eliasson, who has populated museums and cities around the world by publicly demonstrating, through many artful connections between science, technology and ecology, that there are many alternatives to the visual experience of common sense. The art form – or forum – that I chose consisted of asking the three of them to conjoin their forces in presenting films and photographs to set the stage for this famous debate, with Eliasson 'refereeing' the debate through his own work.

It may seem silly to ask an artist to adjudicate a debate between a philosopher and a physicist – especially a debate whose pecking order had been historically settled once and for all: the physicist speaks of the real world, and the philosopher 'does not understand physics'; the artist is irrelevant here. But that was precisely the point, a point shared by Saraceno's heterarchy: that it is now possible to complicate the hierarchy of voices and make the conversation between disciplines move ahead in a way that is more representative of the twenty-first century than of the twentieth. No discipline is the final arbiter of any other.

That is exactly what Elie During did in a brilliant piece of philosophical fiction in which he entirely rewrote the 1922 dialogue as if Einstein had actually paid attention to what Bergson had told him. In the end, 'Zweistein' – that is, the Einstein of 2010 – was not, of course, convinced (that would have been a falsification, and no longer a fiction), but he had to admit that there might be more philosophy in his physics than he had claimed in 1922. So now we have a more balanced situation: the space and time in which we live – experientially, phenomenologically – might not be a mere mistake of our subjective self, but might have some relevance for what the world is really like. Instead of accepting the divide between physics and philosophy, this reenactment was a means of answering Alfred North Whitehead's famous question: 'When red is found in nature, what else is found there also?' Likewise, is it possible to imagine a world where scientific knowledge is able to *add* to the world instead of *dismissing* the experience of being in the world? [...]

Bruno Latour, extracts from 'Some Experiments in Art and Politics', *e-flux journal* (March 2011) [footnotes not included]. (http://eflux.com)

Biographical Notes

Roy Ascott is a British artist and theorist, and Professor of Technoetic Arts at Plymouth University.

Gregory Bateson (1904–80) was a theorist of systems, information, sociology and anthropology.

Mary Catherine Bateson is an independent cultural anthropologist and writer.

Benjamin Bogart is a Canadian artist, media activist, educator and cultural theorist.

Pierre Bourdieu (1930–2002) was a sociologist, ethnographer and cultural theorist.

R. (Richard) Buckminster Fuller (1895–1985) was a systems theorist, architect and designer.

Jack Burnham is a writer and former sculptor who between the 1960s and 1990s was a critic, curator and theorist of art and systems.

Fritjof Capra is a physicist, ecologist and founding director of the Center for Ecoliteracy, Berkeley.

Paolo Cirio is an Italian artist and activist who works across diverse new media fields.

Geoff Cox is Associate Professor, Participatory IT Research Centre, Aarhus University, Denmark.

Jim Crutchfield is Professor of Physics, Complexity Sciences Center, University of California, Davis.

David Dunn is an American composer, sound artist and theorist of acoustic ecology.

Brian Eno is a British composer, musician and artist.

Frank Gillette is an American artist and theorist of video, taxonomic and ecological systems.

Michael Joaquin Grey is an American media artist and technological inventor.

Boris Groys is Global Distinguished Professor of Russian and Slavic Studies at New York University.

Hans Haacke is a German-born American artist based in the US since 1965. He was Professor of Art at the Cooper Union from 1967 to 2002.

Francis Halsall is Lecturer in History and Theory of Modern and Contemporary Art at NCAD, Dublin.

Usman Haque is a designer of interactive architecture systems and founding partner of Umbrellium.

Helen Mayer Harrison and **Newton Harrison** form a practice which fosters collaborations to restore and renew ecological systems and environments. They are Professors Emeriti, University of California, San Diego.

N. Katherine Hayles is Professor of Literature at Duke University.

Caroline A. Jones is Professor of Art History, Massachusetts Institute of Technology.

Stephen Jones is an Australian video artist, technologist, writer and curator.

Christian Katti is a philosopher and writer on art.

Phivos-Angelos Kollias is a Greek composer and writer on musical theory.

Bruno Latour is Professor and Vice-President for Research, Sciences Po, Paris.

Mary Louise Lobsinger is Associate Professor, History and Theory of Architecture, University of Toronto.

Richard Paul Lohse (1902–88) was a Swiss painter, graphic designer and theorist.

James Lovelock is an independent scientist and ecologist.

Alessandro Ludovico is an Italian artist, media critic and editor in chief of *Neural* magazine.

Niklas Luhmann (1927–98) was a sociologist and systems theorist.

Humberto Maturana is a neurobiologist and professor at the University of Chile.

Donella H. Meadows (1941–2001) was an environmental scientist, teacher and writer.

Laurent Mignonneau and **Christa Sommerer** are media artists working in the fields of interactive computer installation and artificial life. They are professors at the Kunstuniversität Linz.

William J. Mitchell (1944–2010) was an architect, urban theorist and professor at MIT (1992–2010).

Manfred Mohr is a German-born digital artist, based in New York since 1981.

Nam June Paik (1932–2006) was a Korean-American video and media artist.

Gordon Pask (1928–96) was a theorist of cybernetics, psychology and educational technology.

Philippe Pasquier is Associate Professor, School of Interactive Arts and Technology, Simon Fraser University.

Zabet Patterson is Assistant Professor, Contemporary Art and Digital Media, Stony Brook University.

Nick Prior is an electronic musician and Head of Sociology at the University of Edinburgh.

Ken Rinaldo is an American artist and educator exploring technological and natural systems interaction. He directs the Art and Technology Program at Ohio State University.

Claude E. Shannon (1916–2001) was a mathematician, electronic engineer and pioneer of information theory.

Sonia Landy Sheridan is an American artist and Professor Emerita of the School of the Art Institute of Chicago.

Christa Sommerer (see Laurent Mignonneau)

UBERMORGEN.COM (lizvlx and Hans Bernhard) is a research-based, interventionist media practice.

Francisco Varela (1946–2001) was a biologist, neuroscientist and theorist.

Steina Vasulka (born in Iceland) and **Woody Vasulka** (born in Brno, now Czech Republic) are experimental artists and pioneers of video art, based in the USA since the early 1960s.

Heinz von Foerster (1911–2002) was a physicist and theorist of cybernetics.

Warren Weaver (1894–1978) was a mathematician, information theorist and computer scientist.

Michael Weinstock is an architect and Director of the Emergent Technologies and Design programme, Architectural Association, London.

Mitchell Whitelaw is Associate Professor of Media Arts and Production, University of Canberra.

Norbert Wiener (1894–1964) was a mathematician and theorist of cybernetics.

Stephen Willats is a British artist and pioneer of socially collaborative practice since the early 1960s.

Iannis Xenakis (1922–2001) was a Greek-French composer, music theorist and architectural engineer.

Bibliography

This section comprises selected further reading and does not repeat the bibliographic references for writings included in the anthology. For these please see the citations at the end of each text.

Apter, Michael J., 'Cybernetics and Art', *Leonardo* (1969) 257–65

Arthur, W. Brian, 'Competing technologies, increasing returns, and lock-in by historical events', *The Economic Journal* (1989) 116–31

Ascott, Roy, *Telematic Embrace: Visionary Theories of Art, Technology and Consciousness*, ed. and intro. Edward A. Shanken (Berkeley and Los Angeles: University of California Press, 2004)

Alloway, Lawrence, *Network: Art and the Complex Present. Contemporary American Art Critics no. 1* (Ann Arbor: UMI Research Press, 1984)

Arnold, Darrell, ed., *Traditions of Systems Theory: Major Figures and Contemporary Developments* (London and New York: Routledge, 2013)

Ashby, W. Ross, *Design for a Brain* (New York: John Wiley, 1952)

Baldwin, Sandy, 'Wiring John Cage: Silence as a Global Sound System', *Electronic Book Review* (1996)

Bateson, Gregory, *Mind and Nature: A Necessary Unity* (New York: Dutton, 1979)

Becker, Howard Saul, *Art Worlds* (Berkeley and Los Angeles: University of California Press)

Beer, Stafford, *Designing Freedom* (London: John Wiley, 1974)

Boden, Margaret A., and Ernest A. Edmonds, 'What is Generative Art?', *Digital Creativity*, vol. 20, no. 1–2 (2009) 21–46

Boulding, Kenneth E., 'General Systems Theory – The Skeleton of Science', *Management Science*, vol. 2, no. 3 (1956) 197–208

Bronfenbrenner, Urie, *Ecological Systems Theory* (London: Jessica Kingsley Publishers, 1992)

Brown, Paul, Charles Gere, Nick Lambert and Catherine Mason, *White Heat Cold Logic; Early British Computer Art 1960–1980* (Cambridge, Massachusetts: The MIT Press, 2009)

Brün, Herbert, 'From Musical Ideas to Computers and Back', *The Computer and Music* (1970) 23–41

Burnham, Jack, *Beyond Modern Sculpture: The Effects of Science and Technology on the Sculpture of This Century* (New York: George Braziller, 1968)

Burnham, Jack, Charles Harper and Judith Benjamin Burnham, *The Structure of Art* (New York: George Braziller, 1971)

Burraston, David M, 'Generative Music and Cellular Automata Bibliography' (http://www.leonardo.info/isast/spec.projects/generativemusic-biblio.pdf)

Burstein, Gabriel, and Constantin Virgil, 'A Kabbalah System Theory of Ontological and Knowledge Engineering for Knowledge Based Systems', *International Journal of Advanced Research in Artificial Intelligence*, vol. 2, no. 2 (2013) (thesai.org)

Capra, Fritjof, *The Web of Life: A New Scientific Understanding of Living Systems* (New York: Random House, 1996)

Checkland, Peter, *Systems Thinking, Systems Practice. Includes a 30-year Retrospective* (London: John Wiley, 1999)

Crutchfield, James P., 'Space-Time Dynamics in Video Feedback', *Physica D: Nonlinear Phenomena*, vol. 10, no. 1–2 (1984) 229–45 (www.journals.elsevier.com)

Crist, Eileen, and H. Bruce Rinker, eds, *Gaia in Turmoil: Climate Change, Biodepletion and Earth Ethics in an Age of Crisis* (Cambridge, Massachusetts: The MIT Press, 2010)

Danto, Arthur C., 'The Artworld', *The Journal of Philosophy* (1964) 571–84

Dubberly, Hugh, Paul Pangaro and Usman Haque, 'What is Interaction?: Are There Different Types?', *Interactions*, vol. 16, no. 1 (2009) 69–75

Dunn, David, 'Nature, Sound Art and the Sacred', *Music from Nature*, ed. D. Rothenberg; *Terra Nova*, vol. 2, no. 3 (1997) 61–71

Dunn, David, Woody and Steina Vasulka, eds, *Eigenwelt Der Apparate-Welt: Pioneers of Electronic Art/ Pioniere der Elektronischen Kunst* (Linz: Ars Electronica, 1992)

Fernández, María, 'Gordon Pask: Cybernetic Polymath', *Leonardo*, vol. 41, no. 2 (2008) 162-168.

Forbes, Angus Graeme, 'Cellular Automata in Image Processing and Geometry Emergence', *Complexity and Computation*, vol. 10 (2014) 253–72

Freidel, David, Linda Schele and Joy Parker, *Maya Cosmos – Three Thousand Years on the Shaman's Path* (New York: William Morrow & Co. Inc., 1993)

Galanter, Philip, 'What is Generative Art? Complexity Theory as a Context for Art Theory', *GA2003 – 6th Generative Art Conference* (2003)

Gell-Mann, Murray, *The Quark and the Jaguar: Adventures in the Simple and the Complex* (London: Macmillan, 1995)

Gere, Charlie, *Art, Time and Technology* (Oxford: Berg, 2006)

Glanville, Ranulph, Albert Muller, Gordon Pask, *Pask Present: An Exhibition of Art and Design Inspired by the Work of Gordon Pask, Cybernetician and Artist* (Vienna: Edition Echoraum, 2008)

Johnson, Steven, *Emergence: The Connected Lives of Ants, Brains, Cities and Software* (New York: Scribner, 2001)

Kahn, Douglas, *Earth Sound Earth Signal: Energies and Earth Magnitude in the Arts* (Berkeley and Los Angeles: University of California Press, 2013)

Krishnamurti, Jiddu, *The Network of Thought* (New York: HarperCollins, 1982)

Langton, Christopher G., 'Studying Artificial Life with Cellular Automata', *Physica D: Nonlinear Phenomena*, vol. 22, no. 1–3 (1986) 120–49 (www.journals.elsevier.com)

Laszlo, Ervin, *The Systems View of the World: A Holistic Vision for Our Time*, 2nd ed. (New York: Hampton Press, 1996)

Lewin, Roger, *Complexity* (Chicago: University of Chicago Press, 2000)

Lichty, Patrick, 'The Cybernetics of Performance and New Media Art', *Leonardo*, vol. 33, no. 5 (2000) 351–4

Lovelock, James, *Gaia: A New Look at Life on Earth* (Oxford: Oxford University Press, 2000)

McKee, Yates, 'The Public Sensoriums of Pulsa: Cybernetic Abstraction and the Biopolitics of Urban Survival', *Art Journal*, vol. 67, no. 3 (2008) 46–67

Macy, Joanna, *Mutual Causality in Buddhism and General Systems Theory: The Dharma of Natural Systems* (Albany: SUNY Press, 1991)

Mallary, Robert, 'Computer Sculpture: Six Levels of Cybernetics', *Artforum* (May 1969) 29–35.

Mathews, Stanley, 'The Fun Palace as Virtual Architecture: Cedric Price and the Practices of Indeterminacy', *Journal of Architectural Education*, vol. 59, no. 3 (2006) 39–48 (www.acsa-arch.org)

Maturana, Humberto R., *Autopoiesis and Cognition: The Realization of the Living* (*Boston Studies in the Philosophy of Science*, no. 42) (New York: Springer, 1980)

Meadows, Donella H., *Thinking in Systems: A Primer* (White River, Vermont: Chelsea Green, 2008)

Medina, Eden, *Cybernetic Revolutionaries: Technology and Politics in Allende's Chile* (Cambridge, Massachusetts: The MIT Press, 2011)

Parikka, Jussi, *Insect Media: An Archaeology of Animals and Technology* (Minneapolis: University of Minnesota Press, 2010)

Parsons, Talcott, *The Social System* (1951) (London and New York: Routledge, 2013)

Pask, Gordon, *The Cybernetics of Human Learning and Performance: A Guide to Theory and Research* (London: Hutchinson Educational, 1975); *Conversation, Cognition and Learning: A Cybernetic Theory and Methodology* (New York: Elsevier, 1975)

Pias, Claus, 'Hollerith "Feathered Crystal": Art, Science and Computing in the Era of Cybernetics', *Grey Room*, no. 29 (2007) 110–34

Pickering, Andrew, *The Cybernetic Brain: Sketches of Another Future* (Chicago: University of Chicago Press, 2010)

Prigogine, Ilya, and Isabelle Stengers, *Order Out of Chaos: Man's New Dialogue with Nature* (New York: Bantam Press, 1984)

Ray, Thomas, 'An Approach to the Synthesis of Life', *Artificial Life* (1992) 371–408

Reichardt, Jasia, *Cybernetic Serendipity: The Computer and the Arts* (London: Institute of Contemporary Arts/Studio International, 1968)

Rosenboom, David, *Extended Musical Interface with the Human Nervous System* (*Leonardo Monograph Series*, no. 1) (San Francisco: International Society for the Arts, Science and Technology, 1990)

Rosenblueth, Arturo, Norbert Wiener and Julian Bigelow, 'Behaviour, Purpose and Teleology', *Philosophy of Science*, vol. 10, no. 1 (1943) 18–24

Sack, Warren, 'Conversation Map: An Interface for Very-Large-Scale Conversations', *Journal of Management Information Systems*, vol. 17, no. 3 (2000) 73–92

Sadler, Simon, 'An Architecture of the Whole', *Journal of Architectural Education*, vol. 61, no. 4 (2008) 108–29

Sassen, Saskia, *The Global City* (Princeton: Princeton University Press, 2001)

Schöffer, Nicholas, ed., Marcel Joray, trans., *Haakon Chevalier* (Neuchatel, Switzerland: Éditions du Griffon, 1963)

Simon, Herbert A., *The Sciences of the Artificial* (1969) (Cambridge, Massachusetts: The MIT Press, 1981)

Sprenger, Florian, and Christina Vagt, guest eds, *The After Lives of Systems*, special issue of *Communication + 1*, vol 3 (2014) (www.communicationplusone.org)

Terranova, Charissa N., 'Systems and Automatisms: Jack Burnham, Stanley Cavell and the Evolution of a Neoliberal Aesthetic', *Leonardo*, vol. 47, no. 1 (2014) 56–62.

Trachtman, Paul, ed., *Cybernetic* (Journal of the American Society for Cybernetics), vol 1, no. 1 (Summer/Fall 1985)

Turner, Fred, *From Cyberculture to Counterculture* (Chicago: Uniersity of Chicago Press, 2006)

van Maanen, Hans, *How to Study Art Worlds: On the Societal Functioning of Aesthetics* (Amsterdam: Amsterdam University Press, 2009)

Varela, Francisco J., Evan Thompson and Eleanor Rosch, *The Embodied Mind: Cognitive Science and Human Experience* (Cambridge, Massachusetts: The MIT Press, 1991)

Vesna, Victoria, ed., *Database Aesthetics: Art in the Age of Information Overflow* (Minneapolis: University of Minnesota Press, 2007)

Volk, Tyler, *Metapatterns: Across Space, Time and Mind* (New York: Columbia University Press, 1995)

von Bertalanffy, Ludwig, *General Systems Theory: Foundations, Development, Applications* (New York: George Braziller, 1968)

von Foerster, Heinz, *Observing Systems* (Seaside, California: Intersystems Publications, 1984)

Wallerstein, Immanuel, 'Culture as the Ideological Battleground of the Modern World-System', *Theory, Culture and Society*, vol. 7, no. 2–3 (1990) 31–55

Weaver, Timothy, 'Biomimetics: Emulation and Propagation in Post-Traditional Ecologies', in Rolf Hughes and Jenny Sundén, eds, *Second Nature: Origins and Originality in Art, Science and New Media* (Stockholm: Axl Books, 2011)

Willats, Stephen, ed., *Control* magazine (1962 to the present) (www.controlmagazine.org)

Youngblood, Gene, *Expanded Cinema* (New York: E.P. Dutton, 1970)

Index

ACKNOWLEDGEMENTS

Editor's acknowledgements

Although I have tried to be inclusive and equitable, this book, like any anthology, is partial and idiosyncratic. Any shortcomings reflect the limits of my own knowledge and the lack of scholarly literature on the topic. This volume would have amounted to much less without the excellent 'nuggets' of feedback and far-reaching discussions generated in the DXARTS Systems seminar that I led at the University of Washington in 2014, which pushed my own thinking about the topic. I am most grateful to Roy Ascott, who introduced me to cybernetics twenty years ago and also to David Dunn and Cristina Cogdell, whose suggestions enriched the book. Special thanks to my research assistant Joel Ong, who made invaluable contributions to all aspects of the project and to Ian Farr, who entrusted the volume to me and nurtured its development with unerring certainty and unwavering support.

For my partners in dancing with systems, Yolande and Jasmin.

Publisher's acknowledgements

Whitechapel Gallery is grateful to all those who gave their generous permission to reproduce the listed material. Every effort has been made to secure all permissions and we apologize for any inadvertent errors or omissions. If notified, we will endeavour to correct these at the earliest opportunity. We would like to express our thanks to all who contributed to the making of this volume, especially Suzanne Anker, Roy Ascott, Mary Catherine Bateson, Hans Bernard, Benjamin Bogart, Jim Crutchfield, Brian Eno, Michael Joaquin Grey, Boris Groys, Hans Haacke, Francis Halsall, Helen Mayer Harrison and Newton Harrison, Usman Haque, N. Katherine Hayles, Amanda Heitler, Caroline A. Jones, Christian Katti, Phivos-Angelos Kollias, Bruno Latour, Mary Louise Lobsinger, The Donella Meadows Institute, Laurent Mignonneau, Manfred Mohr, Nam June Paik Estate, Zabet Patterson, Philippe Pasquier, Emily Pethick, Nick Prior, Ken Rinaldo, Sonia Landy Sheridan, Christa Sommerer, Woody and Steina Vasulka, Michael Weinstock, Mitchell Whitelaw, Stephen Willats, Jane Wolfson. We also gratefully acknowledge the cooperation of Artforum, University of Chicago Press, Paula Cooper Gallery, The Estate of R. Buckminster Fuller, University of Illinois Press, Intellect Books, Richard Paul Lohse Foundation, University of Minnesota Press, MIT Press Journals, Mousse Magazine, Pendragon Press, State University of New York Press, Sage Publications, Shambhala Publications, Stanford University Press.

Whitechapel Gallery

whitechapelgallery.org

Whitechapel Gallery is supported by

ARTS COUNCIL ENGLAND